# Managing Sustainable Business

# Managing Sustainable Business

## Special focus on Information & Communication Technology, and Energy Issues of Sustainable Development Goals

**Sadananda Prusty, Shweta Lalwani, Bright Akwasi Gyamfi and Dipesh Vaya (eds)**

PETER LANG
Chennai - Berlin - Bruxelles - Lausanne - New York - Oxford

Bibliographic information published by the Deutsche Nationalbibliothek.
The German National Library lists this publication in the German National Bibliography; detailed bibliographic data is available on the Internet at http://dnb.d-nb.de.

A catalogue record for this book is available from the British Library.

Library of Congress Cataloging-in-Publication Data

Names: Prusty, Sadananda, 1968- author. | Lalwani, Shweta, 1979- author. | Akwasi Gyamfi, Bright, 1987- author. | Vaya, Dipesh, 1988- author.
Title: Managing sustainable business : special focus on information & communication technology, and energy issues of sustainable development goals / Sadananda Prusty, Shweta Lalwani, Bright Akwasi Gyamfi, Dipesh Vaya.
Description: New York : Peter Lang Publishing, 2025. | Includes bibliographical references and index.
Identifiers: LCCN 2024031367 (print) | LCCN 2024031368 (ebook) | ISBN 9781803745763 (paperback) | ISBN 9781803745770 (ebook) | ISBN 9781803745787 (epub)
Subjects: LCSH: Sustainable development. | Business--Technological innovations.
Classification: LCC HC79.E5 P76 2025 (print) | LCC HC79.E5 (ebook) | DDC 338.9/27--dc23/eng/20240731
LC record available at https://lccn.loc.gov/2024031367
LC ebook record available at https://lccn.loc.gov/2024031368

Cover design by Peter Lang Group AG

ISBN 978-1-80374-576-3 (print)
ISBN 978-1-80374-577-0 (ePDF)
ISBN 978-1-80374-578-7 (ePub)
DOI 10.3726/b22028


Published by Peter Lang Pvt Ltd, Chennai, India
info@peterlang.com - www.peterlang.com



This publication has been peer reviewed.

# Contents

Preface vii

PART I ICT and Business Problems 1

HAYFORD ASARE OBENG, TARIK ATAN, JAPHETH NUHU AHMED, DOMINIC J. KAMARA AND DENNIS AFFUM OSEI

1 Examining the Mediating Role of Patients' Satisfaction and Hospital Reputation on the Influence of Corporate Social Responsibility on Patients' Loyalty in Ghanaian Public Hospitals 3

DAGNU HAILE TESSSEMA, DR. FIGEN YEŞILADA, HAYFORD ASARE OBENG, JAPHETH NUHU AHMED AND DOMINIC J. KAMARA

2 Impact of Total Quality Management on Revisit Intention through the Mediating Role of Patient Satisfaction on Inpatients in Public Hospitals in Addis Ababa, Ethiopia 31

SAMUEL-SOMA M. AJIBADE, OLUWADARE JOSHUA OYEBODE, SUSHOVAN CHAUDHURY AND ANTHONIA OLUWATOSIN ADEDIRAN

3 A Scientometric Analysis of Swarm-based Metaheuristics Approach 51

DR. JAPHETH NUHU AHMED, FIGEN YEŞILADA, HAYFORD ASARE OBENG, DAGNU HAILE TESSEMA, DOMINIC J. KAMARA, DENNIS AFFUM OSEI AND PETER C. MORNYEY

4 The Mediating Effect of Patient Trust on the Link between Relationship Marketing and Hospital Brand Image in Public Hospitals-Nigeria 69

ALEEM ABDUL-KAREEM AND STEPHEN TAIWO ONIFADE
5 The Effect of Organizational Culture on Individual Creativity: An Empirical Investigation of Packaging Firms in Büsan Industrial Site of Central Anatolia Province, Konya-Türkiye 93

PART II ICT, Energy and Financial Sustainability 115

JENEBA JOY TUCKER AND DR. MURAD ABDURAHMAN BEIN
6 The Impact of Credit Risk on Bank's Performance after the World Financial Crises: Evidence from Commercial Bank in Sierra Leone 119

RICHARD SAGOR MITRA AND DR. HEMANT BHANAWAT
7 Analyzes of Common Behavioral Bias among Indo-Arabian Investors and How to Better Manage Them 155

AKASH KUMAR BISWAL AND BISWAJIT PATRA
8 Determinants of Pensions: Empirical Evidence from OECD Countries 173

DR. SUSHMA AND DR. ANURAG SHUKLA
9 Associability between Overconfidence Bias and Financial Risk Tolerance with Reference to North India 201

SHWETA LALWANI
10 Technology Enforcing Upskilling of Workforce for Creating Sustainable Organizations: A Conceptual Review of Trends and Transcending Role of HR 217

List of Contributors 233

Index 237

# Preface

The contemporary business landscape is dynamic, and in light of the escalating ecological predicaments of the current era, adhering to conventional business practices is anticipated to engender further predicaments. Hence, it is imperative to address the escalating environmental issues and the mounting demand for sustainable energy consumption on a global scale. The phenomenon of global warming poses a significant risk to the continued existence of the human species and thus requires appropriate measures to be taken in order to address it effectively. This book is dedicated to exploring strategies for addressing the energy and sustainability crises through the lens of advanced technology, drawing on the trends and experiences of various global economies. In the contemporary business landscape that relies heavily on information and communication technology, a favorable business environment can facilitate the development of an efficient supply chain, promote financial inclusion, and foster effective leadership through a comprehensive economic and financial perspective. Notwithstanding, there exist numerous significant topics to discuss, such as the ecological impact of economic progress in contemporary times, an efficacious commercial milieu, and the patterns of generating renewable and sustainable energy. The topic of sustainability within the context of global supply chains is also addressed in other interconnected areas.

In order for the energy revolution to be successful, it is imperative to have a substantial amount of power derived from renewable sources as well as an efficient, intelligent, decentralized, and secure system for the distribution, storage, and utilization of electricity and heat. The implementation of this measure is expected to have a positive impact on the environment and enhance the overall ease of conducting business operations. This transition has disrupted the energy sector, resulting in significant challenges for the worldwide economy. The establishment of a more robust interconnection among decentralized system components also entails potential hazards. The aforementioned statement holds particularly true in light of the increasing utilization of information and communication technology (ICT) and the mechanization of procedures. On one hand, this phenomenon presents significant prospects for the advancement of novel commercial frameworks, commodities, and methodologies. Conversely, the augmented interconnectivity of disparate components within a decentralized system is not without its potential hazards. The objective is to devise viable solutions for an energy system that is sustainable from economic, environmental, and social perspectives while also satisfying rigorous security requirements.

The development of sustainable energy systems through the application of cutting-edge technologies and the improvement of economic and financial outlooks in a business environment that is heavily reliant on information technology will be the main topics of this research publication. The presentation of innovative and unique concepts may be observed. However, the topics of energy policy, circular economy, life cycle assessment, financial inclusion, enhanced IT systems, effective corporate leadership, and supply chain management could also hold substantial importance. The development of models that can be applied across various temporal and geographical scales may enhance comprehension of the circumstances that facilitate technological and organizational change. It is conceivable to develop novel modeling methodologies that can meet technical and physical constraints while also taking into account economic, environmental, and social aspects.

The urgency to implement effective strategies aimed at addressing energy sustainability arises from the compelling global imperatives posed by climate change, environmental degradation, and resource depletion. Projections indicate a continued escalation in the world's energy demand, primarily propelled by population expansion and economic advancement. Consequently, the imperative for sustainable energy solutions has assumed heightened significance. Notably, fossil fuel combustion persists as the principal contributor to greenhouse gas emissions, thereby significantly impacting climate change dynamics and exacerbating air pollution levels. Absent prompt initiatives to transition towards renewable energy sources and enhance energy efficiency standards, the likelihood of exacerbating these environmental crises looms large, potentially precipitating catastrophic consequences for ecosystems, economies, and human well-being.

Moreover, energy insecurity poses multifaceted social, economic, and geopolitical ramifications, especially in regions heavily reliant on imported fossil fuels or grappling with energy poverty. Dependence on finite and geopolitically precarious energy resources undermines national security, economic resilience, and social equity dynamics. Vulnerable demographic segments, notably those situated in developing countries and marginalized communities, bear a disproportionate burden of energy poverty, lacking access to clean and affordable energy sources essential for sustenance and livelihood activities. Hence, addressing energy sustainability assumes paramount importance, serving as a catalyst for advancing energy access, poverty alleviation efforts, social equity enhancement, and inclusive economic progress.

Furthermore, the transition towards sustainable energy systems heralds unprecedented prospects for fostering innovation, economic expansion, and employment generation. Strategic investments in renewable energy technologies, energy efficiency enhancements, and eco-friendly infrastructure initiatives can catalyze transformative economic growth trajectories, attract substantial investment inflows, and engender the creation of myriad employment opportunities across diverse sectors. By harnessing the potential of clean energy sources such as solar, wind, hydroelectric, and biomass, nations can effectively diversify their energy portfolios, fortify energy security frameworks, and diminish dependence on fossil fuels. Concurrently, the embracement of sustainable energy solutions has the potential to spur technological innovation endeavors, nurture entrepreneurial ventures, and position countries at the vanguard of the global transition towards a low-carbon and resilient future.

PART I

# ICT and Business Problems

Research elucidates previously unknown facts and facilitates the application of theoretical frameworks, thereby contributing to the expansion of knowledge and the development of novel theories. Furthermore, research enables informed decision-making regarding the practical implementation of said theories. However, management research involves the systematic investigation of various dimensions of organizational issues, with the aim of identifying and applying managerial tools and techniques to effectively address these problems. Managerial research involves a thorough, objective, and scientific examination of a problem, with the aim of proposing management procedures that can effectively address the issue at hand. The process involves the methodical acquisition and development of knowledge that is customized to address specific management requirements, with the aim of resolving managerial issues within a designated timeframe. The Rotman School of Management conducts research on a diverse array of managerial topics, utilizing a range of methodologies and drawing from various underlying disciplines. This is an investigation aimed at comprehending the organizational framework and operational mechanisms of management, as well as the methodologies employed in the process of decision-making. Management research can be defined as the systematic investigation of issues pertaining to planning, organizing, directing, and controlling with the aim of addressing problems.

According to these empirical research on ICT and business issues, there are various ways to improve real-world situations and boost corporate operations. The primary focus lies in resolving weaknesses within the ICT infrastructure. This involves strategic investment projects aimed at updating hardware and software systems to ensure smooth communication between different systems, improved operational effectiveness, and robust cybersecurity measures. Moreover, it is crucial to implement extensive training programs to enhance the digital literacy and competency of staff members. Organizations can reduce operational inefficiencies caused by technical limitations by developing a staff that possesses the necessary ICT skills and troubleshooting abilities.

Furthermore, the identified obstacles emphasize the need for regulatory changes and policy structures specifically designed to address and reduce ICT-related barriers in the business environment. These efforts may involve passing laws to promote digital innovation, protect data privacy, and strengthen cybersecurity protocols. Creating a favorable regulatory framework not only encourages investments in ICT infrastructure but also promotes a suitable atmosphere for digital entrepreneurship. Additionally, promoting strategic collaborations and partnerships between public and commercial institutions shows potential in facilitating the exchange of knowledge, pooling of resources, and cooperative actions to tackle common ICT concerns. By synchronizing regulatory frameworks with industrial imperatives and strategically utilizing ICT breakthroughs, firms may successfully overcome obstacles, stimulate technical innovation, and foster sustainable growth models in the digital era.

HAYFORD ASARE OBENG, TARIK ATAN, JAPHETH NUHU AHMED, DOMINIC J. KAMARA AND DENNIS AFFUM OSEI

# 1 Examining the Mediating Role of Patients' Satisfaction and Hospital Reputation on the Influence of Corporate Social Responsibility on Patients' Loyalty in Ghanaian Public Hospitals

ABSTRACT
Organizations have begun placing social responsibility in their core business strategy and are becoming organizations with precepts that may exhibit their social obligations honestly when engaging with customers. In hospital and healthcare services, corporate social responsibility (CRS) is seen as the process of integrating medicine and medical products, ambulance services, nursing services, and food stock by hospitals and healthcare institutions to offer a better life for individuals in society. This study, therefore, seeks to investigate patients' satisfaction and hospital reputation as mediating roles in the association between corporate social responsibility and patient loyalty. The study used purposive and convenient sampling techniques to draw primary data from 422 clients and used AMOS v23 to analyze the data. The study's findings show that CSR significantly and favorably affects patients' loyalty. Also, the results revealed that patients' satisfaction and hospital reputation mediate the interplay between corporate social responsibility and patients' loyalty. The study is important and unique since, in general, CSR activities are implemented by private organizations and mining corporations to address the adverse effects created by their conduct. Therefore, this study put forward the idea that hospital management should see corporate social responsibility as a strategy to foster positive connections with their clients and translate those positive feelings into improved social outcomes for the hospitals.

Keywords: Corporate social responsibility, patient loyalty, patient's' satisfaction, hospital reputation, healthcare institutions, hospital

## Introduction

Businesses and organizations around the world are taking steps to implement plans and strategies that aim at improving the social well-being of individuals and preserving the environment (Armstrong et al., 2023). For instance, the United Nations Global Compact (2014) on sustainability attempts to set its goals and concerns for global businesses and societies to come up with innovative programs and policies that aim at breaking and controlling the impact of corporate social responsibility (CSR). The European Union (2010) refers to CSR as the roles of organizations and their consequences for humanity. The Union underscored the need for enterprises to incorporate ethical consideration, human rights, environmental, and social concerns, as well as co-value creation, into their business strategies in a joint effort with their stakeholders to fully achieve their CSR obligations. The classical perspective of CRS is concerned with philanthropic activities and also addressing social issues, but this has changed in recent times as organizations have shifted their attention to resolving social and environmental problems (Handayani et al., 2017).

Evidence from the Global RepTrak 100 (Reputation Institute, 2016) reveals that CSR is the cornerstone of business strategy and actions and is linked to the significant financial performance of stock value. As a result, CSR is a critical precursor to an organization's reputation. According to Hartmann, Marcos, and Apaolaza (2023) and Carrigan and Attalla (2001), the perceptions of managers and their goals remain a reaction to the increasing prominence that customers offer to socially responsible use and their high beliefs about social interventions. The growing concerns of customers for organizations to perform in a more social and reasonable way have broadly been enhanced through the speedy growth of information technology, which has encouraged a much more related and explicit environment (Markovic et al., 2018; Sierra et al., 2017). Customers and the public are rapidly becoming aware of the poor practices of organizations and thus reprove such organizations whose conduct negatively affects society and the environment (Sharma, 2023; Nyilary et al., 2014) or is insincere (Pope & Weaeras, 2016). Consequently, organizations have begun placing social responsibility in their core business strategy (Sierra et al., 2017) and are becoming organizations with precepts (Iglesias et al., 2020) that may exhibit their social obligations honestly when engaging with customers (Hutchinson et al., 2013).

In hospital and healthcare services, CSR is seen as the process of integrating medicine and medical products, ambulance services, nursing services, and food stock by hospitals and healthcare institutions to offer a better life for individuals in society (Mason, 2023; Andaleeb et al., 2007). Hospitals perform their CSR events in accordance with hospital procedures (Valls Martnez, 2023; Tian et al., 2011). The International Bioethical Committee of UNESCO (2010) reports that hospitals and healthcare facilities can achieve their market and social objectives by relying on general ethical principles and laws. Globally, CSR healthcare services are accessible to patients as different organizations offer assistance in the provision of the healthcare system (Panagiotopoulos, 2023; Hossian & Yahya, 2017). There is a plethora of studies on CSR, but only a few are directed towards the healthcare industry. Furthermore, a greater number of these studies appear to have been carried out in developed countries as compared to developing countries (Kuzey et al., 2021; Deng et al., 2022), such as Ghana.

In Ghana, the government provides the necessary support and budget allocation for various health institutions to implement their programs and projects; for example, the 2018 Budget Statement of the Ministry of Finance offered an approved budget of GH4.42 billion for the health sector. This consisted of funding from the Government of Ghana, the Annual Budget Funding Amount (ABFA), the Internally Generated Fund (IGF), and donor funding. Despite the support from the Government of Ghana and other donor agencies to the health services, the overall improvement of the health and well-being of the population has been slow-moving and far below the expected global target, as there is an average improvement of 50 % as compared to the expected improvement of 75 % (Ghana National Health Policy, 2020). This may be due to the inadequate quality of healthcare services in Ghana (National Healthcare Quality Strategy, 2016) and inequitable access to health services in relation to healthcare facilities and the collection of services offered by the hospitals and facilities, such as rehabilitative, palliative, curative, preventive, and promotional (Ghana National Health Policy, 2020).

Therefore, this study aims to examine the influence of CSR on patients' loyalty (PL), patient satisfaction (PS), and hospital reputation (HR). The study will increase understanding of the management of health institutions in Ghana and the world over and the mechanisms by which organizations can obtain feedback on their CSR initiatives to enhance their reputation and increase patients' satisfaction to obtain and maintain their loyalty. CSR research is significantly more common in advanced economies than in developing or transitioning ones, according to empirical studies. Due to their association with a weak institutional structure that cannot guarantee a consistent supply of social goods, the latter highlights the need for CSR. In this situation, it is expected of organizations to be more concerned with their social objectives to fill the lacuna (Baughn & McItosh, 2007).

The study is important and unique since, in general, CSR activities are implemented by private organizations (Grabner-Kräuter et al., 2023; Fifka, 2012), and mining corporations to address the adverse effects created by their conduct (Ansu-Mensah et al., 2021).

## Concept of Corporate Social Responsibility and Healthcare and Hypothesis Development

While CSR can be traced back to Bowen (2013), who first presented the notion of the social responsibility of corporations in 1950, there is no strong agreement on the term in the current literature (Singh & Kumari, 2023). Carroll (2016) further elucidated several facets of associated duties, such as legal and economic CSR that takes profit maximization into account. The International Organisation for Standardisation (2020) redefines CSR and refers to it as the influence of an organization's choices and actions on society as well as the environment that results in proper conduct and a slide that promotes workable change, consisting of attention given to stakeholders expectations, the health and well-being of society, compliance with extant regulations that are in line with international codes of behavior practice, and is consolidated and implemented in the organization. CSR practices and behaviors are vital tools that ensure the sustainability of the business for value creation, and because all stakeholders contribute to the business' value creation, firms are continuing to share their evidence on CSR practices to gain their positive support, which promotes corporate reputation and enhances investors' confidence for investment judgments (Gu, 2023; Yusoff et al., 2013).

According to Manley et al. (2023) and Oger (2010), pursuant to professional CRS, healthcare organizations should have a collection of values to guide their activities, for example, efficient distribution of resources, equitable access to healthcare, and universal coverage. As a result, implementing social and environmental responsibility can be a critical approach to a hospital's expansion of competitiveness and protection of its public reputation. All stakeholders, as well as the public and wider society, should indeed be taken seriously: non-governmental organizations (NGOs), patients, patient associations, policymakers, the government, healthcare professional regulatory bodies, and the environment. A socially responsible hospital may evaluate ways to carefully deal with waste products that are likely to contaminate the environment (Andeobu et al., 2022; Brandoao et al., 2013).

## CSR and Patients' Loyalty

Customer loyalty has attracted the attention of researchers and marketers in recent times (Nyadzayo & Hajizada, 2016). Huo et al. (2022) suggest that loyalty is a persistent gain, an intellectual activity that results from commitment. According to Mandhachitara and Poolthong (2011), loyalty is a significant ingredient in ensuring the sustainability and long-term growth of a corporation, and studies conducted by Leclercq-Machado et al. (2022) and Karim and Rabiul (2022) confirm this statement. Utilizing socially responsible codes helps brands develop and ensure a continuing relationship with clients (Iglesias et al., 2020). Customers perceive CRS initiatives as elements used in examining the product of an organization that yields trustworthy behaviors (Öberseder,

Schlegelmilch & Murphy, 2013). (CSR) is an essential component of a successful business since it allows for the assessment of consumers' perceptions of brands, which in turn allows for the identification of reasons why customers may rethink their purchase of a brand's product or service (Ben Youssef et al., 2018) and the maintenance of loyal customer relationships (Markovic et al., 2018). As a result, healthcare facilities must consistently engage in CRS activities in an effort to cultivate brand loyalty.

From an empirical viewpoint, scholars have associated CRS with customer loyalty. For example, Pérez and Bosque (2015) and Park et al. (2017) found that CSR is directly connected to clients' opinions of a corporation and that CSR positively influences customers' loyalty. Similarly, CSR has the tendency to increase customers' trustworthiness in a company (Vera-Martínez et al., 2022). Conversely, Salmones et al. (2005) assert that under certain conditions, CSR is capable of reducing patients' appetite for using a firm's product or service. We, therefore, suggest the following hypothesis:

$H_1$: CSR is directly related to Patients' Loyalty.

## Patients' Satisfaction, and Patients' Loyalty

Consumers are the corporation's principal actors, and they are concerned with both its economic and ethical success. Hence, if an organization engages in ethical pursuits like corporate social responsibility programs, its clients are happier and more loyal (Kim, 2022). From a psychological standpoint, satisfaction is viewed as an assessment and comparative process (Maan et al., 2020). Consequently, juxtaposing patients anticipated and real products yields satisfaction. Research suggests that patients' satisfaction is a critical component of the business approach that promotes protracted profitability (e.g., Drosos et al., 2020; Hajar et al., 2020). Isalm et al. (2021) identified a positive relationship between CSR and patients' satisfaction. Also, Leclercq-Machado et al. (2022) argued that the interplay between CSR and customer satisfaction is positive. A greater satisfaction rate with clients affects client loyalty (Manyanga, Makanyeza &

Muranda, 2022). According to Koay, Cheah and Chang (2022), a firm's client loyalty is significantly influenced by its ability to satisfy its customers. Consumers who are happy with the services a business offers are more likely to do business with that company again. Client satisfaction has been employed by researchers as a significant predictor of consumer behavior (e.g., Rather et al., 2019; Islam et al., 2021). CSR thereby affects client satisfaction, which further influences customer loyalty implicitly. From the above assessment, we formulate the following hypotheses:

$H_2$: Corporate social responsibility significantly influences Patients' satisfaction.

$H_3$: Patients' Satisfaction significantly and positively influences Patients' loyalty

$H_4$: Patients' Satisfaction mediates the relationship between CSR and Patients' loyalty.

## Hospital Reputation, Corporate Social Responsibility and Patients' Loyalty

Studies conducted on corporate reputation suggest it is one of the most significant resources of an organization (e.g., Winit, Ekasingh & Sampet, 2023; Kanto, Run & Isa, 2016). Aksak, Ferguson and Duman (2016) posit that managers adapt corporate reputation as a strategy to generate competitiveness for organizations. Porter (2008) describes corporate reputation as a tactical instrument that is employed to attain business strategy purposes. Le (2022) argues that CSR is used as a key strategy to promote the corporate image and meet stakeholders' interests in a time of globalization where firms are operating under high competition.

CSR in hospitals and healthcare institutions is concerned with charitable activities for society to enhance public health. CRS promotes the reputation of hospitals as CSR performance in the hospital increases (Lubis, 2018). However, Phillips et al. (2019) advance that CSR focuses not only on charitable activities; it has developed into an operative marketing instrument used to promote the identity and worth of an organization. Hospitals, in order not to encounter reputational risk that may not only lead to the loss of patients but also legal costs, loss of patients' loyalty, and loss of trust, must share CSR activities with stakeholders. This. This is a vital technique since the damage to reputation might take a number of years to rebuild. Hospitals and healthcare institutions that have comprehensive CSR instruments and records of unprecedented services to society and the environment do not prevent the occurrence of a reputational crisis (Maitlis & Christianson, 2014).

A study conducted by Bahta et al. (2021) on CSR and a company's reputation revealed a significant relationship between a corporation's reputation and CSR activities implemented. Also, the work of Sanie and Durres (2015) on the eminence of corporate CSR suggests a connection between CSR activities and the organization's reputation. Given the challenge of evaluating customer service, reputation plays a crucial role in the health service sector. According to Islam et al. (2021), businesses may boost customer loyalty by improving their reputation. The discussion above thus leads to the following hypotheses:

- $H_5$: CSR practices have a direct effect on hospitals' reputations.
- $H_6$: Hospital reputation influence patients' loyalty
- $H_7$: Hospital reputation indirectly affects the interplay between CSR and Patients' loyalty.

## Methodology

### *Research Design, Sampling Technique and Sample Size*

The study adopted a cross-sectional research design to examine the mediating role of patients' satisfaction and hospital reputation on the relationship between CSR and PL. The study focused on some selected public hospitals in Ghana. Purposive and convenient sampling was used to obtain information from patients of the selected public hospitals in Ghana.

Through communication with the selected hospital's management, purposive and convenient sampling methods were used to sample 422 patients for the study.

Clients of the hospitals were asked whether they would like to participate in this research. The survey questionnaires took the respondents approximately ten minutes to answer, and they then submitted them to management. The survey participants' responses regarding CSR, patient loyalty (PL), patient satisfaction (PS), and hospital reputation (HR) were solicited. Valid responses to a total of 422 questionnaires were collected.

## Measurement

Through a Likert scale with five ranks and pertinent sources, all the variables were measured using modified items from existing scales. The data measured in this study were CSR, PL, PS, and HR. The CSR scale was adapted from the work of Walker and Heere (2011). Additionally, the measure of PL was adapted from the study of Rahman and Saidin (2021). Further, the PS scale was adapted from Ware, Davies-Avery and Stewart (1977) and Nguyen et al. (2021), and finally, measures of HR were adapted from Muthoni and Kinyua (2020).

## Results

### *Reliability and Validity*

The suggested model was analyzed using AMOS v23. As SEM is a multivariate data analysis method often used in the social sciences, we decided

Table 1.1: Reliability test

| Variables | Items | Cronbach's Alpha |
|---|---|---|
| Corporate Social Responsibility (CSR) | 4 | 0.83 |
| Hospital Reputation (HR) | 6 | 0.87 |
| Patients Satisfaction (PS) | 3 | 0.80 |
| Patients Loyalty (PL) | 5 | 0.90 |

to apply it to analyze the proposed model (Hair & Alamer, 2022). We performed CFA to determine the quality of the model fit and then calculated CR, CA, and the average variance extracted (AVE). To put our assumptions to the test, we used the structural model to analyze the correlations among the variables. Cronbach's alpha for the four different measures is depicted in Table 1.1. According to Hair et al. (2017), a CA of 0.70 or higher is preferable. The table shows that all latent variables had Cronbach's alphas above 0.7, indicating that the findings were reliable.

## Confirmatory Factor Analysis

In the present study, we used the CFA approach to evaluate the proposed model. Cronbach's alpha, CR, and AVE were used to evaluate convergent validity (AVE). The threshold for acceptability is a composite reliability index greater than 0.70 (Hair, Ringle & Sarstedt, 2011). All the composite dependability values in Table 1.2 are greater than 0.80, as can be seen. Hence, it was concluded that the statistics were reliable. In addition, AVE estimations were employed to corroborate the reality of convergent validity. AVEs higher than 0.50 are preferred (Ringle, Da Silva & Bido, 2015). All AVE values in the table were greater than 0.50. Hence, it established the existence of convergent validity. As can be seen in Table 1.2, most standardized coefficients are higher than the baseline.

Table 1.2: Confirmatory factor analysis results

| Path | Standard Coefficients | (AVE) | CR |
|---|---|---|---|
| Corporate social responsibility | | | |
| CSR5← HCSR | 0.765 | 0.56 | 0.84 |
| CSR4← HCSR | 0.754 | | |
| CSR3← HCSR | 0.810 | | |
| CSR2← HCSR | 0.650 | | |
| Patients' Loyalty | | 0.71 | 0.93 |
| PL1← PL | 0.802 | | |
| PL2← PL | 0.810 | | |
| PL3← PL | 0.848 | | |
| PL4← PL | 0.792 | | |
| PL5← PT | 0.783 | | |
| Patients' Satisfaction | | | |
| PS3← PS | 0.908 | 0.70 | 0.88 |
| PS2← PS | 0.815 | | |
| PS1← PS | 0.575 | | |
| HR6← HREP | 0.549 | 0.69 | 0.89 |
| HR5← HREP | 0.536 | | |
| HR4← HREP | 0.753 | | |
| HR3← HREP | 0.848 | | |
| HR2← HREP | 0.886 | | |
| HR1← HREP | 0.823 | | |

Also, SEM was used in this work to test the hypotheses. Several metrics, like the chi-square test (22/df) and the RMSEA are used to assess model fitness RMSEA. Hu and Bentler (1999) propose using 0.90 as a cutoff for CFI and TLI and 0.08 as a cutoff for RMSEA.

## Model Goodness of Fit

Several metrics, like the chi-square test (22/df) RMSEA, are used to examine model fitness (RMSEA). Hair et al. (2009) proposed model fit indexes as follows: Chi-square/df (cmin/df) ≤ 2 as good, and ≤ 5 as permissible; CFI ≥ 0.95 as great, ≥ 0.90 as traditional, and ≥ 0.8 as every so often permissible; RMSEA ≤ 0.05 is good, and ≤ 0.08 is moderate; TLI ≥ 0.90, and NFI between 0 & 1 is acceptable. The values for the study's model are given in Table 1.3 and Figure 1.1. The RMSEA value of

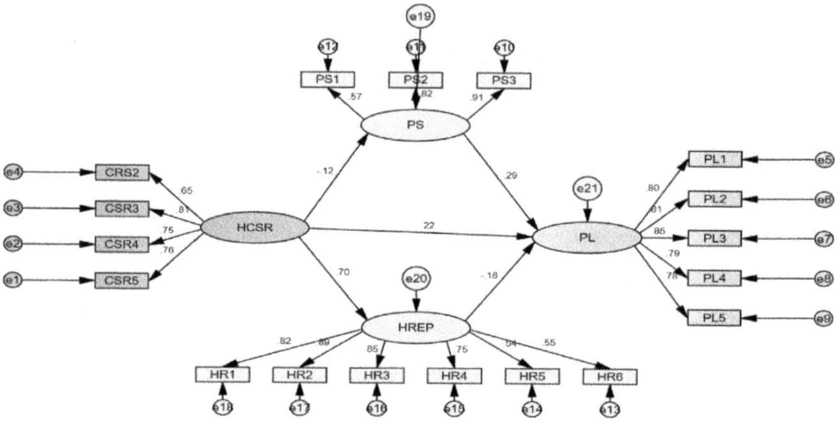


Figure 1.1: AMOS path for CSR effect model

Chi-square = 349.223
Degrees of freedom = 130
Probability level = 0.000

Table 1.3: Hypothesis test result

| Hypothesis | Coefficient | Standard Error | t-value | p-value | Decision |
|---|---|---|---|---|---|
| H1: PL←HCSR | 0.219 | 0.084 | 2.654 | 0.008 | Accepted |
| H2: PS ← HCSR | −0.120 | 0.063 | −2.182 | 0.029 | Accepted |
| H3: PL ← PS | 0.293 | 0.048 | 5.404 | *** | Accepted |
| H5: HREP ← HCSR | 0.697 | 0.058 | 9.506 | *** | Accepted |
| H6: HREP← PL | −0.179 | 0.103 | −2.235 | 0.025 | Accepted |

0.062, which is within the acceptable range, showed a good fit between the model and the empirical covariance matrix. The NFL, TLI, and CFI scores were 0.921, 0.939, and 0.948, respectively, which are all greater than 0.9, and the Chi-square (CMIN/DF ratio) of 2.686 demonstrated that the study's model was accepted.

## Hypothesis Testing

Table 1.3 displays the findings of the predicted value and testing hypotheses of the parameters.

From Table 1.3, a t-value of 2.654 and a p-value of 0.008 with a corresponding effect size of 0.219 indicate that CSR has a direct and significant effect on patients' loyalty to the Ghanaian healthcare service. This means that whenever a hospital or healthcare service improves its CRS activities, patients' loyalty is more likely to increase by 21.9 %. Therefore, H1 suggests that CSR directly influences patients' loyalty is accepted. The analysis also shows a significant but negative effect of CSR on patients' satisfaction. This was revealed by the t-value of −2.182, p-value of 0.029 and effect size of −0.120. We therefore, accept H2, which suggests that corporate social responsibility significantly influences patients' satisfaction. The test result further demonstrated that patients' satisfaction is significant and has positive influence on patients' loyalty with a t-value of 5.404, p-value of 0.000 and a coefficient of 0.293. This implies that patient satisfaction has a tendency of increasing patients' loyalty by 29.3 % whenever there is an improvement in the services of CSR implementation. We, therefore, accept H3, which suggests that patients' satisfaction significantly and positively influences their loyalty. Moreso, the analysis findings demonstrated that CSR has a strong and favorable influence on hospitals' reputations, with a p-value of 0.000, t-value of 9.506 and an effect size of 0.697. The implication is that there is a possibility that hospital reputation increases by 69.7 % whenever CSR activities are undertaken. We therefore accept H5, which suggests that CSR practices have a direct effect on hospitals' reputations. Finally, as shown by the negative t-value value of 2.235, p-value of

0.025, and negative effect size of 0.179, statistical testing demonstrates that HREP has a negative and significant impact on patients' loyalty. Despite the statistical significance of the link, the negative sign shows that HREP has no direct effect on patients' loyalty, indicating that patients would stick with their hospitals regardless of their reputation. Therefore, we accept H6 practices that suggest that hospital reputation influences patients' loyalty.

## Result of Mediating Effect

In order to assess the indirect effect of both intervening variables, patients' satisfaction and hospital reputation, on the association between CSR activities and patients' loyalty, H4 and H7 were tested with the use of the bootstrapping technique. To analyze the impact of the mediating variables, one of the most effective and reliable interpolation approaches is the bootstrapping technique (Zhu et al., 2022).

Table 1.4 gives a snapshot of how CSR indirectly influences patient loyalty by acting as a mediator between patients' satisfaction and the hospital's reputation. The outcome of the findings demonstrated a total negative effect size of −0.162, p-value of 0.005 which is less than 0.05. This outcome suggests that patients' satisfaction and hospital reputation partially mediate the connection between HCSR and patients' loyalty. Hence, we accept H4 and H7, which suggest that patients' satisfaction mediates the interplay between CSR and patients' loyalty and that hospital reputation mediates the CSR and patients' loyalty nexus, respectively. Table 1.5 provides a summary of the direct, indirect, and total effects of HCSR, PS, and HREP on PL.

Table 1.4: Indirect effect results

| Hypothesis | Effect Size | p-value | Decision |
|---|---|---|---|
| H4:H7: PL ←PS←HREP←HCSR | −0.162 | 0.005 | Accepted |

Table 1.5: Direct, indirect and total effect

| | | HCSR | PS | HREP | PL |
|---|---|---|---|---|---|
| Direct Effect | PS | −0.251* | | | 0.000 |
| | HREP | 0.429** | | | 0.000 |
| | PL | 0.032* | 0.151** | −0.470* | 0.000 |
| | | | | | |
| Indirect Effect | PS | 0.000 | 0.000 | 0.000 | 0.000 |
| | HREP | 0.000 | 0.000 | 0.000 | 0.000 |
| | PL | −0.303** | 0.000 | 0.000 | 0.000 |
| | | | | | |
| Total Effect | PS | −0.251* | 0.000 | 0.000 | 0.000 |
| | HREP | 0.429** | 0.000 | 0.000 | 0.000 |
| | PL | −0.053 | 0.151 | −0.470 | 0.000 |

The indirect (mediated) effect of HCSR on PL is significantly different from zero at the 0.01 level ($p = 0.005$ two-tailed). And direct is significant at $p = 0.01$ and 0.05. This is a bootstrap approximation obtained by constructing two-sided bias-corrected confidence intervals.

## Discussion

The impacts of CSR on behavioral outcomes were analyzed, along with the underlying mechanisms. In the healthcare business, where it is taken for granted that healthcare staff already provide social activities and assistance to the communities they serve, CSR is still in its infant stage in developing nations compared to the West. Theoretically, CSR is an important framework for gaining an edge in the market by increasing customer loyalty to a company. To test the hypothesized indirect impacts of CSR on loyalty in the Ghanaian health care sector, we add mediation strategies to our proposed model. The results of the current study corroborated the assumptions, proving that CSR is an effective method for boosting organizational favorability and standing in the eyes of the public. The findings further elucidate the desirable and significant impact of hospital CSR on patients' loyalty. This is in tandem with the findings of Perez and Bosque (2015) and Park et al. (2017), who found that CSR is directly related to clients' opinions of a corporation and that CSR positively influences customers' loyalty, respectively.

The study's finding contradicts the findings of Salmones et al. (2005), who assert that under certain conditions, CSR is capable of reducing patients' appetite for using a firm's product or service. Additionally, the findings demonstrated a significant and positive effect of patients' satisfaction on patients' loyalty. This supports the findings of Manyanga et al. (2022), who suggest that a greater satisfaction rate with clients affects client loyalty, and the study of Koay et al. (2022), who advance that a firm's client loyalty is significantly influenced by its ability to satisfy its customers. Further, the study's findings identified a significant positive effect of hospital CSR on HR, and this confirms the works of Sanie and Durres (2015) on the eminence of corporate CSR activities and the organization's reputation and Islam et al. (2021), who identified that corporate reputation has a direct link with customer loyalty. The final facet of the study revealed that patients' satisfaction and hospital reputation mediate the relationship between hospitals' CSR and patients' loyalty. The verifiable outcomes align with our presumptions and confirm past studies such as Islam et al. (2021). According to the results, CSR is a powerful tool for improving patients' impressions of hospitals and healthcare services and for giving institutions a forum in which to boast about going above and beyond their legal obligations to the communities in which they operate.

## Theoretical Implication

This study makes a variety of conceptual additions to the existing literature. To begin with, this research fills a lacuna in the literature by confirming the analytical model of what influences customers to remain loyal in the setting of a developing nation. According to the research reviewed, this factor plays a crucial role in determining hospital clients' final behavior. CSR is also playing a major role in the framework of a sustainable economy that is used to conducting itself ethically. Less is known regarding how CSR affects developing nations, particularly with regard to CSR and responsible behavior on the part of healthcare. To the best of our knowledge, this study broadens the existing literature on how ethical efforts in hospitals are linked to favorable behavioral outcomes and ultimately lead to a strong corporate reputation. Again, the effect of CSR on the loyalty of customers in the healthcare industry has been studied indirectly by a few researchers who have neglected to account for the significance of corporate competencies. As previous studies have shown, CSR is essential for encouraging customer loyalty, even at the expense of a company's actual competence. Further, the research offers empirical evidence for the suggested concept that CSR impacts consumer loyalty in the service industry through mediating and moderating mechanisms. Lastly, the current study offers recommendations for business and government leaders in emerging markets.

## Managerial Implication

The management ramifications of our study are substantial. Firstly, our research confirms that CSR activities are a component people evaluate when selecting healthcare providers; therefore, our results may serve as suggestions for upper executives. Patients are savvier and more enlightened than at any time in history about the services of hospitals, and our results show they will reward hospital facilities that demonstrate CSR initiatives. Hence, hospital management should see CSR as a tactic for fostering positive connections with their clients and translating those positive feelings into improved social outcomes for the hospitals. Hospital administrations should spread the news about these events and programs via extensive planning that capitalizes on the power of good word of mouth and corporate identity. Those in charge of a hospital will see a change in patient behavior if they don't implement CSR efforts. Additionally, the analysis of this study proposes that business leaders can improve their company's image, customer happiness, and loyalty by catering to the factors that influence customers' actions. According to the findings of this study, it is advised that the healthcare industry invest in ethical efforts, as patients tend to show their support for and preference for those facilities that are regarded as socially responsible by incorporating the greatest loyalty to them. Hospitals and healthcare institutions in poor nations should prioritize spending on CSR because of the positive effect it has on loyalty via a process that is not directly visible. This study's findings suggest that healthcare providers should prioritize ethical initiatives since people are more likely to be loyal to establishments that they see as socially responsible. Due to the good influence CSR has on loyalty via a process that is not readily evident, hospitals and healthcare institutions in developing countries should prioritize expenditure on CSR. Further, the results of this research show that CSR has a substantial impact on the behavioral outcomes of consumers, suggesting that telecom firms should allocate resources towards developing this area. Companies in the telecom industry would be wise to increase their spending on corporate social responsibility (CSR) initiatives since they are very successful at gaining the satisfaction and confidence of their clientele. The company's legitimacy, authenticity, and distinctiveness among rivals may all improve as a consequence of money spent on CSR operations. Finally, the results of this investigation suggest that CSR spending is correlated with positive reactions from customers. A common misconception about corporate social responsibility is that hospitals' CSR efforts only improve patient satisfaction (Ahmad et al., 2022). Evidence from the past supports the idea that corporate social responsibility (CSR) activity favorably impacts business (Gu, 2023) and client outcomes (Park et al., 2017). Hence, corporate social responsibility (CSR) strategies should be enhanced and implemented to establish a sustainable competitive advantage. Yet, targeted CSR and a hospital's emphasis on corporate responsibility are excellent ways to boost patient satisfaction, reputation, and dedication.

## Limitation of the Study

The current study is without limitations. First, we employed purposive and convenient sampling to obtain data from patients who access healthcare from Ghana's hospital facilities. Future studies should consider employing different sampling strategies with a larger sample. Second, since our results are founded on a cross-sectional design, further longitudinal research is necessary to confirm the association between CSR and client outcomes as well as the validity of our findings. Client reactions to the social practices the hospitals engage in may be examined through longitudinal research, where senior management dedication and large resources are indispensable in this regard. Further research was done in Ghana. For improved data generalization, future studies can be conducted in other West African nations. The study once again concentrated only on public hospitals in Ghana. Future researchers can take into account Ghana's public and private healthcare institutions to improve data generalization. Future studies can also consider hospital value and patients' trust as mediating or moderating variables to determine the indirect and strong effects that CSR may have on patients' loyalty.

## Conclusion

This study provided empirical support for the idea that CSR affects patients' loyalty in the healthcare sector through hidden processes. The current study offers empirical proof, supported by conceptual justification, that CSR is the best way to please current clients who later develop a sense of loyalty. Moreover, CSR is a reliable tactic to improve hospitals' reputations. The research revealed that patient loyalty is primarily influenced by two factors: patient satisfaction and hospital reputation. The connection between CSR and social and financial performance has been studied in earlier literature. The present study, on the other hand, revealed the research and put out a conceptual framework to look at CSR in the milieu of clients. Furthermore, the findings demonstrate that patient satisfaction and hospital reputation influenced the association between CSR and patients' loyalty. As a result, this study closes a significant research lacuna by suggesting the client multi-mediation concept and emphasizing the client context. Future studies and decision-makers can greatly benefit from this research's vital knowledge. The framework offers a comprehensive examination of the factors that determine and mediate patients' loyalty, and it confirms how senior management in the environment of emerging nations could use CSR as a viable method to accomplish their goals.

## Editors Note

Examining the mediating role of patients' satisfaction and hospital reputation on the influence of corporate social responsibility on patients' loyalty in Ghanaian public hospitals is crucial for sustainability in several ways. Firstly, by understanding how corporate social responsibility (CSR) initiatives impact patients' loyalty through satisfaction and reputation, hospitals can tailor their CSR strategies to maximize positive outcomes and foster long-term relationships with patients. This, in turn, contribute to the sustainability of public healthcare systems by ensuring patient retention, promoting trust in healthcare institutions, and enhancing overall patient experiences. Additionally, by identifying the mediating effects of satisfaction and reputation, hospitals can pinpoint areas for improvement in their CSR efforts, leading to more effective resource allocation and enhanced social impact. Ultimately, this research provides valuable insights for policymakers, hospital administrators, and healthcare practitioners in Ghana to develop sustainable CSR practices that benefit both patients and the broader healthcare ecosystem.

## References

Ahmad, N., Ullah, Z., AlDhaen, E., Han, H., Araya-Castillo, L., & Ariza-Montes, A. (2022). Fostering hotel-employee creativity through micro-level corporate social responsibility: A social identity theory perspective. *Frontiers in psychology*, *13*, 1–12.

Aksak, E. O., Ferguson, M. A., & Duman, S. A. (2016). Corporate social responsibility and CSR fit as predictors of corporate reputation: A global perspective. *Public Relations Review*, *42*(1), 79–81.

Andaleeb, S. S., Siddiqui, N., & Khandakar, S. (2007). Patient satisfaction with health services in Bangladesh. *Health Policy and Planning*, *22*(4), 263–273.

Andeobu, L., Wibowo, S., & Grandhi, S. (2022). Medical waste from COVID-19 pandemic—a systematic review of management and environmental impacts in Australia. *International Journal of Environmental Research and Public Health*, *19*(3), 1381.

Ansu-Mensah, P., Marfo, E. O., Awuah, L. S., & Amoako, K. O. (2021). Corporate social responsibility and stakeholder engagement in Ghana's mining sector: A case study of Newmont Ahafo mines. *International Journal of Corporate Social Responsibility*, *6*(1), 1–22.

Armstrong, F., Wyns, A., Colagiuri, P., Anderson, R., Hunter, A., Arabena, K., … Borda, A. (2023). Healthy, regenerative and just: Guiding the development of a national strategy on climate, health and well-being for Australia. *The Journal of Climate Change and Health*, *10*, 100205.

Bahta, D., Yun, J., Islam, M. R., & Bikanyi, K. J. (2021). How does CSR enhance the financial performance of SMEs? The mediating role of firm reputation. *Economic Research-Ekonomska Istraživanja*, *34*(1), 1428–1451.

Baughn, C. C., & McIntosh, J. C. (2007). Corporate social and environmental responsibility in Asian countries and other geographical regions. *Corporate Social Responsibility and Environmental Management*, *14*, 189–205. <https://doi.org/10.1002/(ISSN)15353966>

Ben Youssef, K., Leicht, T., Pellicelli, M., & Kitchen, P. J. (2018). The importance of corporate social responsibility (CSR) for branding and business success in small and medium-sized enterprises (SME) in a business-to-distributor (B2D) context. *Journal of Strategic Marketing*, *26*(8), 723–739.

Bowen, H. R. (2013). *Social Responsibilities of the Businessman*. University of Iowa Press. <https://doi.org/10.2307/j.ctt20q1w8f>

Brandao, C., Rego, G., Duart, I., & Nunes, R. (2013). Social responsibility: A new paradigm of hospital governance? *Health Care Analysis*, *21*(4), 390–402.

Carrigan, M., & Attalla, A. (2001). The myth of the ethical consumer—do ethics matter in purchase behaviour. *Journal of Consumer Marketing*, *18*(7), 560–578.

Carroll, A. B. (2016). Carroll's pyramid of CSR: Taking another look. *International Journal of Corporate Social Responsibility*, *1*(1), 1–8.

CSR Europe. (2010). *A guide to CSR in Europe: Country insights by CSR Europe's National Partner Organization (CSR Europe)*. Brussels: CSR Europe.

Deng, Y., Cherian, J., Ahmad, N., Scholz, M., & Samad, S. (2022). Conceptualizing the role of target-specific environmental transformational leadership between corporate social responsibility and pro-environmental behaviors of hospital employees. *International Journal of Environmental Research and Public Health, 19*(6), 3565.

Drosos, D., Kyriakopoulos, G. L., Arabatzis, G., & Tsotsolas, N. (2020). Evaluating customer satisfaction in energy markets using a multicriteria method: The case of electricity market in Greece. *Sustainability, 12*(9), 3862.

European Commission. (2010). *Corporate Social Responsibility (CSR)*. <https://ec.europa.eu/info/business-economy-euro/doing-business-eu/corporate social-responsibility-csr_en> [Accessed on February 12, 2023].

Fifka, M. (2012). The development and state of research on social and environmental reporting in global comparison. *Journal für Betriebswirtschaft, 62*, 45–84.

Ghana National Healthcare Quality Strategy. (2016). *Report on the strategy, coordination and accountability framework.* <https://www.moh.gov.gh/wp-content/uploads/2017/06/National20Quality20Strategy20Ghana.pdf> [Accessed on January 12, 2023].

Ghana National Healthcare Quality Strategy. (2020). *Ensuring healthy lives for all.* <https://www.moh.gov.gh/wp-content/uploads/2020/07/NHP_12.07.2020.pdf-13072020-FINAL.pdf> [Accessed on November 5, 2024].

Grabner-Kräuter, S., Tafolli, F., & Breitenecker, R. J. (2023). Consequences of public sector employees' CSR perceptions in a developing country: Organizational benefits and beyond. *Public Performance & Management Review, 46*(2) 1–40.

Gu, S. (2023). Corporate social responsibility and customer loyalty: The mediating role of co-creation and customer trust. *Asian Journal of Economics, Business and Accounting, 23*(3), 1–19.

Hair, J., & Alamer, A. (2022). Partial Least Squares Structural Equation Modeling (PLS-SEM) in second language and education research: Guidelines using an applied example. *Research Methods in Applied Linguistics, 1*(3), 100027.

Hair, J., Ringle, C. and Sarstedt, M. (2011). PLS-SEM: Indeed a Silver Bullet. *Journal of Marketing Theory and Practice, 19*, 139–151. <https://doi.org/10.2753/MTP1069-6679190202>

Hair, J. F., Hult, G. T. M., Ringle, C. M. and Sarstedt, M. (2017). *A Primer on Partial Least Squares Structural Equation Modeling (PLS-SEM)* (2nd edn). Thousand Oaks, CA: Sage.

Hair Jr., J. F., Black, W. C., Babin, B. J. and Anderson, R. E. (2009). *Multivariate Data Analysis* (7th edn). Saddle River, NJ: Prentice Hall.

Hair Jr, J. F., & Sarstedt, M. (2019). Factors versus composites: Guidelines for choosing the right structural equation modeling method. *Project Management Journal, 50*(6), 619–624.

Hajar, M. A., Ibrahim, D. N., Darun, M. R., & Al-Sharafi, M. (2020). Improving Mobile value-added services (MVAS) in Yemen mobile companies. *International Journal of Information Technology*, *4*(2), 1–11.
Handayani, R., Wahyudi, S., & Suharnomo, S. (2017). The effects of corporate social responsibility on manufacturing industry performance: The mediating role of social collaboration and green innovation. Business: *Theory and Practice, 18*, 152–159.
Hartmann, P., Marcos, A., & Apaolaza, V. (2023). Past, present, and future of research on corporate social responsibility advertising. *International Journal of Advertising*, *42*(1), 87–95.
Hossain, M. S., & Yahya, S. B. (2017). Developing conceptual framework of corporate community development practices of Islamic banks in Bangladesh: An institutional perspective. *The Journal of Internet Banking and Commerce*, *22*(1), 1–20.
Hu, L. T., & Bentler, P. M. (1999). Cutoff criteria for fit indexes in covariance structure analysis: Conventional criteria versus new alternatives. *Structural Equation Modeling: A Multidisciplinary Journal*, *6*(1), 1–55.
Huo, C., Hameed, J., Zhang, M., Bin Mohd Ali, A. F., & Amri Nik Hashim, N. A. (2022). Modeling the impact of corporate social responsibility on sustainable purchase intentions: Insights into brand trust and brand loyalty. *Economic Research-Ekonomska Istraživanja*, *35*(1), 4710–4739.
Hutchinson, B., Singh, D., Svensson, G., & Mysen, T. (2013). Towards a model of conscientious corporate brands: A Canadian study. *Journal of Business & Industrial Marketing*, *28*(8), 687–695.
Iglesias, O., Markovic, S., Bagherzadeh, M., & Singh, J. J. (2020). Co-creation: A key link between corporate social responsibility, customer trust, and customer loyalty. *Journal of Business Ethics*, *163*, 151–166.
The International Organization for Standardization. (2020). *Corporate Social Responsibility (CSR)—Definition, history and evolution*. <https://youmatter.world/en/definition/csr-definition/> [Accessed on January 14, 2023].
Islam, T., Islam, R., Pitafi, A. H., Xiaobei, L., Rehmani, M., Irfan, M., & Mubarak, M. S. (2021). The impact of corporate social responsibility on customer loyalty: The mediating role of corporate reputation, customer satisfaction, and trust. *Sustainable Production and Consumption*, *25*, 123–135.
Kanto, D. S., de Run, E. C., & bin Md Isa, A. H. (2016). The reputation quotient as a corporate reputation measurement in the Malaysian banking industry: A confirmatory factor analysis. *Procedia-Social and Behavioral Sciences*, *219*, 409–415.
Karim, R. A., & Rabiul, M. K. (2022). The relationships of corporate sustainability, customer loyalty, and word of mouth: The mediating role of corporate

image and customer satisfaction. *Journal of Quality Assurance in Hospitality & Tourism, 25*(3) 1–21.

Kim, R. C. (2022). Rethinking corporate social responsibility under contemporary capitalism: Five ways to reinvent CSR. *Business Ethics, the Environment & Responsibility, 31*(2), 346–362.

Koay, K. Y., Cheah, C. W., & Chang, Y. X. (2022). A model of online food delivery service quality, customer satisfaction and customer loyalty: A combination of PLS-SEM and NCA approaches. *British Food Journal, 124,* 4516–4532.

Kuzey, C., Uyar, A., Nizaeva, M., & Karaman, A. S. (2021). CSR performance and firm performance in the tourism, healthcare, and financial sectors: Do metrics and CSR committees matter?. *Journal of Cleaner Production, 319*, 128802.

Le, T. T. (2022). Corporate social responsibility and SMEs' performance: Mediating role of corporate image, corporate reputation and customer loyalty. *International Journal of Emerging Markets 18,* 4565–4590.

Leclercq-Machado, L., Alvarez-Risco, A., Esquerre-Botton, S., Almanza-Cruz, C., de las Mercedes Anderson-Seminario, M., Del-Aguila-Arcentales, S., & Yáñez, J. A. (2022). Effect of Corporate social responsibility on consumer satisfaction and consumer loyalty of private banking companies in Peru. *Sustainability, 14*(15), 9078.

Lubis, A. N. (2018). Corporate social responsibility in health sector: A case study in the government hospitals in Medan, Indonesia. *Verslas: teorija ir praktika, 19*(1), 25–36.

Maan, A. T., Abid, G., Butt, T. H., Ashfaq, F., & Ahmed, S. (2020). Perceived organizational support and job satisfaction: A moderated mediation model of proactive personality and psychological empowerment. *Future Business Journal, 6*, 1–12.

Maitlis, S., & Christianson, M. (2014). Sensemaking in organizations: Taking stock and moving forward. *Academy of Management Annals, 8*(1), 57–125.

Mandhachitara, R., & Poolthong, Y. (2011). A model of customer loyalty and corporate social responsibility. *Journal of Services Marketing 25*(2), 122–133.

Manley, J., Garner, C., Halliday, E., Lee, J., Mattinson, L., Mckeown, M., … Wood, J. (2023). Saving lives and minds: Understanding social value and the role of anchor institutions in supporting community and public health before and after COVID-19. In *Corporate social responsibility in the health sector: CSR and COVID-19 in global health service institutions* (pp. 193–217). Cham: Springer International Publishing.

Manyanga, W., Makanyeza, C., & Muranda, Z. (2022). The effect of customer experience, customer satisfaction and word of mouth intention on customer loyalty: The moderating role of consumer demographics. *Cogent Business & Management, 9*(1), 2082015.

Markovic, S., Iglesias, O., Singh, J. J., & Sierra, V. (2018). How does the perceived ethicality of corporate services brands influence loyalty and positive word-of-mouth? Analyzing the roles of empathy, affective commitment, and perceived quality. *Journal of Business Ethics, 148*, 721–740.

Mason, A., Spencer, E., Barnett, K., & Bouchie, J. (2023). Examining the prominence and congruence of organizational corporate social responsibility (CSR) communication in medical tourism provider websites. *Journal of Hospitality and Tourism Insights, 6*(1), 1–17.

Ministry of Health. (2020). *National health policy: Ensuring healthy lives for all* (Rev. edn). Ghana. <https://www.moh.gov.gh/wp-content/uploads/2020/07/NHP_12.07.2020.pdf-13072020-FINAL.pdf> [Accessed on January 12, 2023].

Muthoni, D. M., & Kinyua, G. M. (2020). Corporate reputation and firm performance: An empirical analysis of motor vehicle assemblers in Nairobi city county, Kenya. *Journal of Business and Economic Development, 5*(2), 73–81.

Nguyen, N. X., Tran, K., & Nguyen, T. A. (2021). Impact of service quality on in-patients' satisfaction, perceived value, and customer loyalty: A mixed-methods study from a developing country. *Patient Preference and Adherence, 15*, 2523–2538.

Nyadzayo, M. W., & Khajehzadeh, S. (2016). The antecedents of customer loyalty: Moderated mediation model of customer relationship management quality and brand image. *Journal of Retailing and Consumer Services, 30*, 262–270.

Nyilasy, G., Gangadharbatla, H., & Paladino, A. (2014). Perceived greenwashing: The interactive effects of green advertising and corporate environmental performance on consumer reactions. *Journal of Business Ethics, 125*(4), 693–707.

Öberseder, M., Schlegelmilch, B. B., & Murphy, P. E. (2013). CSR practices and consumer perceptions. *Journal of Business Research, 66*(10), 1839–1851.

Oger, C. (2010). Corporate social responsibility in the pharmaceutical industry: Between trend and necessity, *Journal of Literature Pharmacy Sciences, 12*(1), 82–91. Toronto: Toronto Metropolitan University.

Panagiotopoulos, Φ. I. (2023). Corporate social responsibility initiatives and programs in the health system of Greece due to the pandemic of COVID-19. In *Corporate social responsibility in the health sector: CSR and COVID-19 in global health service institutions* (pp. 93–110). Cham: Springer International Publishing.

Park, E., Kim, K. J., & Kwon, S. J. (2017). Corporate social responsibility as a determinant of consumer loyalty: An examination of ethical standard, satisfaction, and trust. *Journal of Business Research, 76*, 8–13.

Pérez, A., & Rodríguez del Bosque, I. (2015). How customer support for corporate social responsibility influences the image of companies: Evidence from the banking industry. *Corporate Social Responsibility and Environmental Management, 22*(3), 155–168.

Phillips, S., Thai, V. V., & Halim, Z. (2019). Airline value chain capabilities and CSR performance: The connection between CSR leadership and CSR culture with CSR performance, customer satisfaction and financial performance. *The Asian Journal of Shipping and Logistics, 35*(1), 30–40.

Pope, S., & Wæraas, A. (2016). CSR-washing is rare: A conceptual framework, literature review, and critique. *Journal of Business Ethics, 137*(1), 173–193.

Porter, M. E. (2008). *Competitive strategy: Techniques for analyzing industries and competitors.* New York, NY: Simon and Schuster.

Rahman, W. A. W. A., & Saidin, Z. H. (2021). Customer loyalty: The effect of service quality and brand Image in Malaysias automotive industry. *International Journal of Asian Social Science, 11*(4), 188–199.

Rather, R. A., Tehseen, S., Itoo, M. H., & Parrey, S. H. (2019). Customer brand identification, affective commitment, customer satisfaction, and brand trust as antecedents of customer behavioral intention of loyalty: An empirical study in the hospitality sector. *Journal of Global Scholars of Marketing Science, 29*(2), 196–217.

Reputation Institute. (2016). *Global RepTrak 100 report.* Retrieved from <https://www.pwc.com/gx/en/ceo-survey/2016/landing-page/pwc-19th-annual-global-ceosurvey>. [Accessed February 12, 2023].

Ringle, C., Da Silva, D., & Bido, D. (2015). Structural equation modeling with the SmartPLS. Bido, D., da Silva, D., & Ringle, C. (2014). Structural equation modeling with the Smartpls. *Brazilian Journal of Marketing, 13*(2), 56–73.

Salmones, M. D. M. G. D. L., Crespo, A. H., & Bosque, I. R. D. (2005). Influence of corporate social responsibility on loyalty and valuation of services. *Journal of business ethics, 61*, 369–385.

Sanie, D. O. D. A., & Durres, A. (2015). The importance of corporate social responsibility. *Journal of Sociological Research, 6*(1), 86–91.

Sharma, A., Choudhury, M., Agarwal, S., Sharma, R., & Sharma, R. (2023). Corporate social responsibility and roles of developers for sustainability in companies. In *The route towards global sustainability: Challenges and management practices* (pp. 313–332). Cham: Springer International Publishing.

Sierra, V., Iglesias, O., Markovic, S., & Singh, J. J. (2017). Does ethical image build equity in corporate services brands? The influence of customer perceived ethicality on affect, perceived quality, and equity. *Journal of Business Ethics, 144*(3), 661–676.

Singh, K., & Kumari, P. (2023). CSR: Analysing shifting paradigm from climate change to global health emergencies. *World Review of Science, Technology and Sustainable Development, 19*(1–2), 55–69.

Tian, Z., Wang, R., & Yang, W. (2011). Consumer responses to corporate social responsibility (CSR) in China. *Journal of Business Ethics, 101*, 197–212.

UNESCO. (2010). *Report of the International Bioethics Committee of UNESCO on social responsibility and health.* <https://en.unesco.org/themes/ethics-science-and-technology/ibc/reports-and-advices> [Accessed on January 15, 2023].

United Nations Environment Programme. (2014). *United Nations Global Compact—Environmental principles training package.* Available from <www.unep.fr/shared/publications/other/.../en/UNGC TRAIN ALL.pdf> [Accessed February 15, 2023].

Valls Martínez, M. D. C., Soriano Román, R., Grasso, M. S., & Martín-Cervantes, P. A. (2023). Corporate social responsibility and profitability in Spanish private health care during the COVID-19 period. In *Corporate social responsibility in the health sector: CSR and COVID-19 in global health service institutions* (pp. 173–192). Cham: Springer International Publishing.

Vera-Martínez, J., Alvarado-Herrera, A., & Currás-Pérez, R. (2022). Do consumers really care about aspects of corporate social responsibility when developing attitudes toward a brand?. *Journal of Global Marketing, 35*(3), 193–207.

Walker, M., & Heere, B. (2011). Consumer attitudes toward responsible entities in sport (CARES): Scale development and model testing. *Sport Management Review, 14*, 153–166.

Ware Jr, J. E., Davies-Avery, A., & Stewart, A. L. (1977). *The measurement and meaning of patient satisfaction.*

Winit, W., Ekasingh, E., & Sampet, J. (2023). How disclosure types of sustainability performance impact consumers' relationship quality and firm reputation. *Sustainability, 15*(1), 803.

Yusoff, H., Mohamad, S. S., & Darus, F. (2013). The influence of CSR disclosure structure on corporate financial performance: Evidence from stakeholders' perspectives. *Procedia Economics and Finance, 7*, 213–220.

Zhu, Z., Liu, B., Yu, Z., & Cao, J. (2022). Effects of the digital economy on carbon emissions: Evidence from China. *International Journal of Environmental Research and Public Health, 19*(15), 9450.

DAGNU HAILE TESSSEMA, DR. FIGEN YEŞILADA,
HAYFORD ASARE OBENG, JAPHETH NUHU AHMED
AND DOMINIC J. KAMARA

# 2 Impact of Total Quality Management on Revisit Intention through the Mediating Role of Patient Satisfaction on Inpatients in Public Hospitals in Addis Ababa, Ethiopia

ABSTRACT

Purpose: Research regarding the application of total quality management in the healthcare industry of developing countries, mainly Africa, is underestimated. There are a few papers on the issue related to the healthcare sector. Thus, the study objective is to study the impact of total quality management (TQM) on patient revisit intention (PRI) through the mediating role of patient satisfaction (PSAT) on in-patients in Ethiopian public healthcare industry.

Methods: The questionnaire was used through convenience sampling, and the approach to data analysis was quantitative. The sample size is 300 in-patients who were being treated in state hospitals in the capital city Addis Ababa, Ethiopia. Data were collected from July to September 2022, and they were analyzed using Amoss 26 and SPSS 26.

Findings: Total quality management had a significant and positive impact on PRI through the mediating role of PSAT. Similarly, TQM has a positive impact on PSAT, and PSAT has a positive impact on PRI, WOM, and PTR. Besides this, PTR has a positive impact on WOM and PRI. WOM also has a positive impact on PRI.

Limitations: The research was conducted in hospitals, so to generalize the result of this research; it needs to be replicated in other service industries. Secondly, the information was gathered from in-patients in state hospitals in Addis Ababa. Future scholars can investigate by engaging with both inpatients and outpatients to acquire a comprehensive understanding of healthcare services.

Implications: This paper has a significant contribution to existing literature and further research since there is a limited number of research regarding the application of TQM in the healthcare sector compared to manufacturing sectors. This study has also made a vital contribution to managers in healthcare organizations' strategic planning and decision-making.

Keywords: Patient trust, patient satisfaction, word-of-mouth, revisit intention, TQM, healthcare sector

## Introduction

TQM is one of the management systems that enhance continuous service or product improvement to have better efficiency and effectiveness, through which firm performance is improved to have patient satisfaction. Birhanu et al. (2010) acknowledged that the foundation of the quality of medical treatment in the twenty-first century is the conviction that in Ethiopia, long-term relationships between medical professionals and their patients are the most effective way to give the best possible care.

According to Sabella et al. (2014), the introduction of the TQM system, which has generated a large amount of interest among academics and practitioners, has resulted in a large body of literature. W. Edwards Deming led the quality revolution in the late 1970s, with major assistance from Crosby and Juran (Bisgaard, 2008). TQM also plays a significant role in increasing service quality, which is the main concern of the health sector for patient satisfaction. In addition, the quality of hospital-patient relationships and patient loyalty were positively impacted by physicians and hospital managers who supported organizational management systems. Prior research focused on TQM in the manufacturing sector, neglecting the expanding attention on the quality of healthcare services and its consequences (Agyapong, Afi, & Kwateng, 2018). The only few studies that have been found have concentrated on developed economies as well as a few developing economies in Asia and Eastern Europe. The emerging African economies have not received much attention. Furthermore, current research has frequently concentrated up on direct link among patient satisfaction and patient revisit intention without much attention to the potential relations between TQM and revisit intention, which satisfaction might mediate.

## Literature Review

### *Total Quality Management*

According to Abbas (2020), TQM significantly increases individual capacities to innovate new goods or improve the performance of existing products or services by emphasizing training, development, and continuous improvement in all areas. The application of TQM is also necessary for the employees in the health sector to do their work effectively and increase the retention of nurses and doctors. As emphasized by Alshrbaji et al. (2022), most previous research has focused on TQM application mainly in the manufacturing industry, and less TQM research has been done in developing and Arab countries compared to Western countries, ignoring other industries such as healthcare. Researchers and gurus, especially Deming, Crosby, and Juran, have highly contributed to the development of TQM with the Fourteen Points of Deming, continuous improvement, and zero defects, respectively (Mohammed et al., 2020).

Aktar (2021) suggested that the quality of treatment provided in healthcare institutions is frequently overlooked in medical research and rarely investigated scientifically. As a result, the objective of this research is to analyze the impact of total quality management on PRI through the mediation role of PSAT. Interactions with healthcare providers is crucial for patient revisit intention. Total quality management is a tool for service quality through continuous improvement, through which patient satisfaction is formed (Lashgari et al., 2015). The application of TQM in the healthcare industry can amplify the effectiveness and efficiency of doctors, nurses, and all health professionals through proper enlightenment about continues improvement in their work place, performance of employees and communication with patients. In recent research by Nguyen, Tran, and Nguyen (2021), a positive relationship between TQM and patient satisfaction was established.

## Patient Satisfaction

According to Kebede et al. (2021), despite the fact that most hospital workers are involved in traditional medicine distribution and inventory management, this conventional method has an impact on patient satisfaction and pharmacists' potential role in undertaking responsibilities in the healthcare industry of Ethiopia. Numerous works on patient satisfaction and healthcare service delivery have conducted. However, there are a limited number of researchers, particularly patients, who are satisfied with healthcare services in Ethiopia. The healthcare sector needs to strive for patient satisfaction since it is prominent for the well-being of society, which is vital for assessing the healthcare industry. As emphasized by Yilma et al. (2021), research findings in Ethiopia revealed a variety of issues related to patient satisfaction. Researchers have identified different reasons that lead to unhappy patients, such as ineffective communication with doctors and nurses and a lack of free treatment. Some scholars have found a positive and direct association between customer satisfaction, patient trust, and revisits (Han & Ryu, 2009; Oliver et al., 1997). As shown by Ganesan (1994), service satisfaction has a positive impact on doctor-patient trust, through which patients recommend the institute to other patients through WOM and revisit healthcare organizations.

## Patient Trust

As emphasized by Isa et al. (2019), patients are more likely to have a long-term relationship with the facility and are more inclined to return if they believe hospital staff is honest and trustworthy, and they are more likely to revisit the hospital if they are willing to recommend it to their friends and family.

According to Liu et al. (2021), patient trust is an essential antecedent of patient intent to revisit and patient satisfaction. Patient satisfaction is the degree to which consumers feel that their expectations are met after purchasing goods or using services. It affects patient trust positively, and additionally, it fosters patient loyalty towards the healthcare provider. The primary focus of the current research was an early investigation of patient-doctor relationship trust, particularly associated with the idea that the doctors' words and deeds are trustworthy. In regard to the relationship between trust and satisfaction, Shabbir et al. (2010) found that the level of trust has a favorable two-directional impact on satisfaction, and trust developed has a direct and positive impact on subsequent satisfaction, which is two-directional. Trust still has a direct connection to word-of-mouth (WOM), despite the fact that it does not significantly mediate a two-directional relationship between WOM and the patient's readiness to return. Gambarov et al. (2017) found a positive and significant relationship between patient loyalty and trust in the healthcare sector. Patient trust also was found to have mediating effect between patient satisfaction and WOM (Ganesan, 1994).

## Word-of-Mouth

Based on the study of Sweeney et al. (2008), the ability of WOM to influence views or behaviors is dependent on the sender-receiver relationship and the message's depth and intensity. WOM has been recognized as a main source of informational influence on consumer repurchase decision-making as well as a channel for conveying happiness or discontent with a service encounter (Nguyen et al., 2020; Robinson et al., 2020). Positive and negative word-of-mouth are both possible. Positive WOM occurs when a customer promotes a product or service to others, while negative WOM occurs when a customer complains to others about a product or service. As claimed by Zhou et al. (2017), the tendency of a patient to return to the same healthcare professional is referred to as patient revisit intention. Only a few studies have verified the beneficial association between word-of-mouth and the intention to return to the healthcare service (Cham et al., 2016).

## Patient Revisit Intention

Based on the research of Afridi et al. (2020), consumers who disseminate favorable, positive WOM are essential for the sector's competitive advantage. Loyal clients are seen as the foundation of success for the sector, as they improve a company's performance and image. Positive WOM and revisit intention are used to gauge customer loyalty, implying that loyal patients urge the health sector to be efficient and innovative. According to Sholeh and Chalidyanto (2021), as a consequence of the use of comprehensive quality management concepts, patient satisfaction with the hospital's service quality has an effect on increasing the degree of loyalty. The healthcare industry should give priority to patients' safety and deliver service with efficiency and effectiveness. TQM had influence on PRI through the mediating role of patient satisfaction. Patients who are satisfied with the quality of the service and communication between doctors and medical staff revisit the hospital. As the researcher is aware, there is research to be done about the impact of TQM on patient revisit intention through mediation role of PSAT in healthcare organizations in developing countries, mainly in Africa. The review of the related literatures leads to the following eight hypotheses:

H1: Total Quality Management has a significant and positive impact on patient satisfaction.

H2: PSAT has a significant and positive impact on PRI.

H3: PSAT has a significant and positive impact on PTR.

H4: PSAT has a significant and positive impact on WOM.

H5: PTR has a significant and positive impact on WOM.

H6: PTR has a significant and positive impact on the PRI.

H7: WOM has a significant and positive impact on PRI.

H8: TQM has a significant and positive impact on PRI through the mediating role of PSAT

# Methodology

### *Research Design and Data Collection*

The samples in this study were in-patients in state hospitals in Addis Ababa, Ethiopia. The sampling technique used was convenience sampling of in-patients, both male and female, whose ages were over 18 years. The sample size was 300 in-patients in selected state hospitals. Questionnaire was distributed for data collection. TQM, PSAT, PTR, WOM, and PRI were the variables in this research. A quantitative research and cross-sectional approach are applied in this study to assess the effect of TQM on PRI through the mediation of PSAT.

In this research paper, a Likert scale with a five-scale interval is used, namely "1 = strongly disagree" and "5 = strongly agree" for all construct items except patient satisfaction, which is used, namely "1 = strongly dissatisfied" and "5= strongly satisfied." An Ethiopian translator converted the questionnaires used for the purpose of this research from English to Amharic. The Amharic version was interpreted back into English by a professional translator to assess the translation's trustworthiness.

In general, a single item is too limited to encompass all of a construct's characteristics, and it is also likely to have a high rate of measurement error (Churchill, 1979). Thus, three measurement items to measure total quality management were adapted from Alshrbaji et al. (2022) and Nguyen and Nagase (2019). Five items were adapted from Siripipatthanakul (2021) to measure WOM. The five-item measure of patient satisfaction is adapted from Nguyen et al. (2021). A four-item scale for revisit intention is adapted from Qin and Prybutok (2013) and Gambarov et al. (2017). A five-item scale for patient trust is adapted from Gambarov (2017) and Han and Hyun (2015).

## Data Analysis Method

To test the research hypothesis, the collected data were first entered into SPSS and then explored with AMOS 26. Individual elements were categorized, and a construct was established before the statistical investigation of the hypotheses. SPSS and Amos software were utilized to assess the study framework. Cronbach's alpha, AVE and CR were examined. According to Forza and Filippini (1998), in order to apply the SEM method, the recommended sample size need to 5–10 times rule and observation 50–400. The sample size for this study was 300 which meet the requirement.

## Results and Discussion

### *Reliability and Validity*

The items in the questionaries related to TQM are process quality (four items, PQ1–PQ4), interaction quality (five items, IQ–IQ5), and environment quality (three items, EQ1–EQ3). The items for patient satisfaction (five items, PSAT1-PSAT5), patient trust (five items, PTR1–PTR5), word-of-mouth (five items, WOM1–WOM5), and patient revisit intention (PRI1–PRI5) Cronbach`s alpha was used to measure the reliability of the items. As shown in Table 2.1, the alpha values for TQM items ranged from 0.848 to 0.883; PSAT was 0.881; PRI was 0.861; PTR was 0.883; and WOM was 0.890. The value of Cronbach`s alpha for the reliability test was greater than 0.70, which indicates that constructs are internally consistent Hair et al. (2010) TQM items (PQ2, IQ3, and IQ5), WOM1, and PSAT5 are deleted while analyzing the model fit using modification indices.

Table 2.1: Cronbach's alpha value

| **Items** | **Items (Before Delated)** | **Cronbach's Alpha Value (Before Delated)** | **Items after Delated** | **Cronbach's Alpha Value after Delated** |
|---|---|---|---|---|
| TQM | | | | |
| PQ | 4 | 0.891 | 3 | 0.877 |
| IQ | 5 | 0.903 | 3 | 0.848 |
| EQ | 3 | 0.883 | 3 | 0.883 |
| PSAT | 5 | 0.901 | 4 | 0.881 |
| PTR | 5 | 0.883 | 5 | 0.883 |
| WOM | 5 | 0.878 | 4 | 0.890 |
| PRI | 4 | 0.861 | 4 | 0.861 |

## Confirmatory Factor Analysis (CFA)

Reliability and validity of the measurement model was examined. The model fit was also assessed, Chi-square/DF (CMIN/DF) (<3 good, <5 acceptable), the comparative index (CFI): >0.95 excellent, >0.90 acceptable, and the RMSEA value (<0.08 acceptable) (Hair et al., 2010). To enhance the model's fit, a modification index requirement <10 was looked at. Standardized regression weights >0.5 were utilized to create the model (Hair et al., 2009).

Table 2.2 indicates the coefficients >0.5, the value of CR is between 0.794 and 0.881 greater than 0.7, and AVE is 0.526–0.598 greater than 0.5, which are in the accepted range as stated by Hair et al. (2009).

Table 2.2: Analyze of CFA

| Path | Standardized Coefficients | Cronbach's Alpha | AVE | CR |
|---|---|---|---|---|
| *Total Quality Management (TQM)* | | | | |
| TQM→PQ1 | 0.57 | 0.867 | 0.565 | 0.794 |
| TQM→PQ3 | 0.55 | | | |
| TQM→IQ1 | 0.53 | | | |
| TQM→IQ2 | 0.55 | | | |
| TQM→IQ4 | 0.54 | | | |
| TQM→EQ1 | 0.67 | | | |
| TQM→EQ2 | 0.68 | | | |
| TQM→EQ3 | 0.54 | | | |
| TQM→PQ4 | 0.60 | | | |
| *Patient Satisfaction (PSAT)* | | | | |
| PSAT→SAT1 | 0.80 | 0.881 | 0.534 | 0.821 |
| PSAT→SAT2 | 0.82 | | | |
| PSAT→SAT3 | 0.82 | | | |
| PSAT→SAT4 | 0.78 | | | |
| *Patient Trust (PTR)* | | | | |
| PTR→PTR1 | 0.72 | 0.883 | 0.598 | 0.881 |
| PTR→PTR2 | 0.74 | | | |
| PTR→PTR3 | 0.77 | | | |
| PTR→PTR4 | 0.85 | | | |
| PTR→PTR5 | 0.75 | | | |
| *Word-of- Mouth (WOM)* | | | | |
| WOM→WOM2 | 0.84 | 0.890 | 0.526 | 0.815 |
| WOM→WOM3 | 0.89 | | | |
| WOM→WOM4 | 0.66 | | | |
| WOM→WOM5 | 0.75 | | | |
| Patient Revisit Intention (PRI) | | | | |
| PRI→PRI1 | 0.81 | 0.861 | 0.586 | 0.846 |
| PRI→PRI2 | 0.77 | | | |
| PRI→PRI3 | 0.80 | | | |
| PRI→PRI4 | 0.75 | | | |

Table 2.3: The goodness-of-fit measures of the model

| **Goodness of Fit** | **Measure** | **Score** | **Threshold** | **Conclusion** |
|---|---|---|---|---|
| Absolute fit | Chi-square<br>DF<br>GFI<br>Chi-square/<br>DF<br>RMSEA | 518.033<br>279<br>0.882<br>1.857<br>0.054 | <br><br>>0.85<br>Less than 5<br>0.05–0.08 | <br><br>Accepted |
| Incremental fit | NFI<br>TLI<br>CFI | 0.900<br>0.942<br>0.951 | Between 0 & 1<br>>0.90<br>>0.90 | Accepted |
| Parsimony fit | AGFI | 0.851 | >0.80 | Accepted |

## Model Goodness of Fit

The model fit has acceptable values, as indicated in Table 2.3. The RMSEA value of 0.054 is less than 0.08 and the GFI 0.882 is within an acceptable range of greater than 0.85. The incremental fit value is also within the acceptable range. The value of Chi-square/DF was 1.857, which is also acceptable. A parsimony fit was also performed, and an AGFI value of 0.851 is accepted.

## Hypotheses Testing

Based on the results obtained from Amos and model fit, the hypothesis is tested (Table 2.4), and the outputs are shown in Table 2.4. The result of two side significance tests is that the first seven hypotheses of the study are accepted.

Table 2.4: Structural model results

| Hypoth-esis | Path | β-Value | S.E. | t- Statistics | P-Value | Results |
|---|---|---|---|---|---|---|
| H1 | TQM→PSAT | 0.69 | 0.115 | 7.391 | *** | Supported |
| H2 | PSAT→PTR | 0.45 | 0.061 | 6.548 | *** | Supported |
| H3 | PSAT→PRI | 0.35 | 0.060 | 5.410 | *** | Supported |
| H4 | PSAT→WOM | 0.23 | 0.071 | 3.520 | *** | Supported |
| H5 | PTR→WOM | 0.44 | 0.090 | 6.060 | *** | Supported |
| H6 | PTR→PRI | 0.18 | 0.073 | 2.631 | 0.009 | Supported |
| H7 | WOM→PRI | 0.34 | 0.056 | 5.150 | *** | Supported |

The result of the hypothesis test indicates that TQM→PSAT is supported (β = 0.69, t-value =7.391, P = 0.001). The finding consistent with previous research (Nguyen & Nagase, 2019; Alshrbaji et al., 2022).

The path coefficient of PSAT→PTR was statistically significant (β = 0.45, t-value = 6.548, P = 0.001). The result implies that satisfied patients will trust the healthcare provider and revisit. Others also found that PSAT has a positive impact on PTR (Hung et al., 2019; Liu et al., 2021). The standardized coefficient of PSAT→PRI was 0.35 and statistically significant (β = 5.410, t-value = 5.410, P = 0.001). PSAT has a positive impact on PRI, which is in accordance with the findings of Siripipatthanakul (2021) and Nguyen et al. (2021).

The standardized coefficient of PSAT→WOM was positive (0.23) and statistically significant (β = 0.23, t-value = 3.520, P = 0.001). The finding is consistent with the previous work of Nguyen et al. (2021), Siripipatthanakul (2021), and Salim and Dewi (2018). The result implies that PSAT is the main determinant of WOM in the medical service. Patients who are satisfied with the perceived services of the hospital can deliver positive WOM to friends and family members about the healthcare industry or hospital. The standardized coefficient of PTR→WOM was 0.44, which was statistically significant (β = 044, t-value = 6.060, P = 0.001), indicating the positive influence of PTR on word-of-mouth, consistent with the findings of Salim and Dewi (2018). Patients who trust the healthcare organization can spread positive word of mouth.

The path PTR→PRI is also statistically significant (β = 0.18, t-value = 2.631, P = 0.01). The result is consistent with the findings of Kurnianingrum and Hidayat (2020) and Mohd Isa et al. (2019). The direct effect of WOM on patient revisit intention (WOM→PRI) was statistically significant (β = 0.34, t-value = 5.150, P = 0.001). The finding is consistent with the finding of Mohd Isa et al. (2019).

## Mediation Analysis

The path TQM-PSAT-PRI shows that there is an indirect effect of TQM on patient revisit intention (β = 0.445, P = 0.002), as given in Table 2.5.

This suggests that total quality management influences patient revisit intention favorably through the mediation of patient satisfaction. When the health sector uses TQM concepts to improve waiting times, communication, and quality, patients are inclined to return to the same healthcare organization because they are happy with the care they received. A patient's intention to return increases by 0.456 when TQM increases by one standard deviation (P = 0.01), and accordingly, the eighth hypothesis was also supported. Since no researchers have examined a comparison of overall TQM within the healthcare industry and the intention to revisit the healthcare sector, particularly in developing nations, this study and hypothesis add fresh information to the body of literature.

The structure model shows that WOM has a positive impact on revisit intention, with a standardized coefficient of 0.34. This indicates that when patients are satisfied with the services provided by the healthcare

Table 2.5: Indirect effect

| **Hypothesis** | **Path** | **Coefficient (β)** | **P-Value** | **Result** |
|---|---|---|---|---|
| H8 | TQM→PSAT→PRI | 0.456 | 0.002 | Supported |

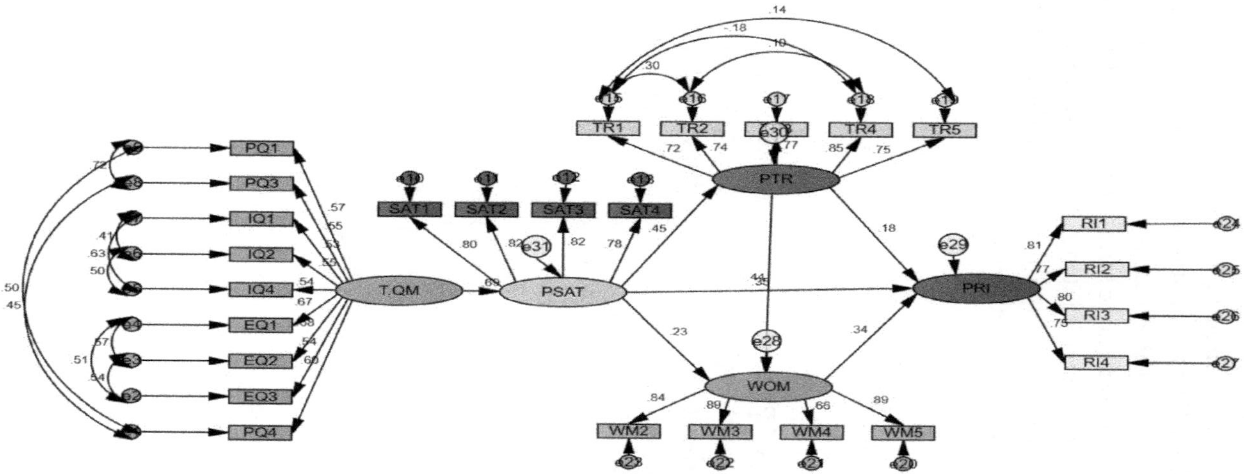


Figure 2.1: Results of the structural model

industry, they are more likely to spread good word of mouth, which increases the likelihood that they will return. With a standardized coefficient of 0.18, patient trust has a positive effect on PRI as well, indicating that patients who trusted the hospital or doctor were more likely to return. Also, patients who have confidence in the hospital will recommend it to their friends and family. Besides this, TQM has a significant and positive impact on PSAT with a standardized coefficient of 0.69, and PTR has a lower positive impact on PRI.

## Conclusion

In our study, the effects of TQM on PSAT, patient satisfaction on RI, patient trust on WOM, and PTR on PRI were assessed. Positive forces have been identified. Total quality management is important for the healthcare industry by enhancing efficiency and effectiveness in responding to or delivering service to patients and enabling patients to revisit the hospital or healthcare provider. TQM adds value to the company and makes it more competitive. The hypotheses used for this research paper were supported. The research's conclusions have important contributions to make to the hospital or health industry through improved service-delivering systems, mainly through the application of TQM principles. The services offered by hospitals have an effect on PSAT, which subsequently contributes to PRI through the mediation of PSAT. To increase customer happiness, the service organization should take these aspects into account in its strategic planning.

## Limitation and Future Research Direction

Although the study's objectives were met and it made a significant contribution to knowledge and literature, there were some drawbacks. First, in order to generalize the findings of this study, researchers can consider other geographic areas since this study was limited to Ethiopia. Second, the data were collected from inpatients at state hospitals in Addis Ababa. Future researchers might investigate both inpatients and outpatients to acquire a comprehensive understanding of healthcare service quality. Third, although the data were gathered from patients who were admitted, managers and staff from the healthcare sector may be included in future studies. Fourth, future research must take into account both public and private healthcare service providers. Finally, the proposed study ought to assess the direct impact of TQM on PTR, WOM, and PRI. In addition to this, since this study is cross-sectional, future studies can concentrate on longitudinal research.

## Theoretical Implication

This paper has a significant contribution to existing literature and further research since there is limited research on TQM in service sector (healthcare) compared to manufacturing. In addition to this, there is a larger research gap in developing countries than in Western or developed countries, so researchers can concentrate on this area of research to increase the number of limited publications. Also, this study has important applications for healthcare administrators who may desire to take the aspects covered in this paper into account and use them when making strategic plans and decisions. This study has a significant impact on managers' ability to increase patient satisfaction through the use of total quality management concepts and to perform better in a competitive and changing environment. These elements should be taken into account by managers in the health industry when designing service quality programs, and it is advised that hospitals implement TQM techniques for patient satisfaction and revisit intention.

## Editors Note

Exploring the impact of Total Quality Management (TQM) on revisit intention through the mediating role of patient satisfaction among in-patients in public hospitals in Addis Ababa, Ethiopia, holds profound implications for sustainability within the healthcare sector. By elucidating the mechanisms through which TQM practices influence patient satisfaction and subsequent revisit intentions, this research provide valuable insights into improving service quality and patient outcomes. Enhanced patient satisfaction resulting from effective TQM implementation not only fosters trust and loyalty among patients but also contributes to the long-term viability and resilience of public healthcare institutions. Moreover, by identifying the mediating role of patient satisfaction, this study can inform strategic decision-making processes aimed at optimizing resource allocation and enhancing service delivery practices to ensure the continued provision of high-quality care. Ultimately, this research has the potential to drive positive organizational change, promote sustainable healthcare practices, and foster a culture of continuous improvement within Addis Ababa's public hospital system.

## References

Abbas, J. (2020). Impact of total quality management on corporate green performance through the mediating role of corporate social responsibility. *Journal of Cleaner Production, 242*, 118458. <https://doi.org/10.1016/j.jclepro.2019.118458>.

Afridi, S. A., Haider, M., & Shahjehan, A. (2020). Customers' loyalty through healthcare quality; mediating role of customers' commitment: A comparative study of public and private hospitals. *City University Research Journal, 10*(1), 27–46.

Agyapong, A., Afi, J. D., & Kwateng, K. O. (2018). Examining the effect of perceived service quality of health care delivery in Ghana on behavioural intentions of patients: The mediating role of customer satisfaction. *International Journal of Healthcare Management, 11*(4): 276–288. <https://doi.org/10.1080/20479700.2017.1326703>.

Aktar, M. S. (2021). Determinates service quality and its effect on patients' satisfaction of private medical college hospitals, Rangpur, Bangladesh. *International Journal of Management and Accounting, 3*(4), 91–105. <https://doi.org/10.34104/ijma.021.0910105>.

Alshrbaji, M., Mohammed, M., & Shamayleh, A. (2022). The impact of total quality management and perceived service quality on patient satisfaction in healthcare: A systematic review. In *2022 advances in science and engineering technology international conferences, ASET 2022* (April). <https://doi.org/10.1109/ASET53988.2022.9734872>.

Birhanu, Z., Assefa, T., Woldie, M., & Morankar, S. (2010). Determinants of satisfaction with health care provider interactions at health centres in Central Ethiopia: A cross sectional study. *BMC Health Services Research,* 10. <https://doi.org/10.1186/1472-6963-10-78>.

Bisgaard, S. (2008). Quality management and Juran's legacy. *Quality Engineering, 20*(4), 390–401. <https://doi.org/10.1080/08982110802317398>.

Cham, T. H., Lim, Y. M., Aik, N. C., & Tay, A. G. M. (2016). Antecedents of hospital brand image and the relationships with medical tourist' behavioral intention. *International Journal of Pharmaceutical and Healthcare Marketing,* 10(4), 412–431. <https://doi.org/10.1108/IJPHM-02-2016-0012>.

Churchill, G. A. (1979). A paradigm for developing better measures of marketing constructs a paradigm for developing better measures of marketing constructs. *Journal of Marketing Research,* 16(1), 64–73. <https://doi.org/10.2307/3150876>.

Forza, C., and Filippini, R. (1998). TQM impact on quality conformance and customer satisfaction: A causal model. *International Journal of Production Economics, 55*(1), 1–20. <https://doi.org/10.1016/S0925-5273(98)00007-3>.

Gambarov, V., Sarno, D., Hysa, X., Calabrese, M., & Bilotta, A. (2017). The role of loyalty programs in healthcare service ecosystems. *TQM Journal, 29*(6), 899–919. <https://doi.org/10.1108/TQM-02-2017-0019>.

Ganesan, S. (1994). Determinants of long-term orientation in buyer-seller relationships. *Journal of Marketing, 58*(2), 1–19. <https://doi.org/10.1177/002224299405800201>.

Hair Jr., J. F., Black, W. C., Babin, B. J. and Anderson, R. E. (2009). *Multivariate Data Analysis* (7th edn). Saddle River, NJ: Prentice Hall.

Hair, J. F., Black, W. C., & Babin, B. J. (2010). *Multivariate Data Analysis: A Global Perspective* (7th edn). Saddle River, NJ: Pearson.

Han, H., & Hyun, S. S. (2015). Customer retention in the medical tourism industry: Impact of quality, satisfaction, trust, and price reasonableness. *Tourism Management, 46*, 20–29. <https://doi.org/10.1016/j.tourman.2014.06.003>.

Han, H., & Ryu, K. (2009). The roles of the physical environment, price perception, and customer satisfaction in determining customer loyalty in the restaurant industry. *Journal of Hospitality and Tourism Research, 33*(4), 487–510. <https://doi.org/10.1177/1096348009344212>.

Hung, S. W., Cheng, M. J., & Chiu, P. C. (2019). Do antecedents of trust and satisfaction promote consumer loyalty in physical and virtual stores? A Multichannel view. *Service Business, 13*(1), 1–23. <https://doi.org/10.1007/s11628-018-0364-y>.

Kebede, H., Tsehay, T., Necho, M., & Zenebe, Y. (2021). Patient satisfaction towards outpatient pharmacy services and associated factors at Dessie Town Public Hospitals, South Wollo, North-East Ethiopia. *Patient Preference and Adherence, 15*, 87–97. <https://doi.org/10.2147/PPA.S287948>.

Kurnianingrum, A. F., & Hidayat, A. (2020). The influence of service quality and price perception on consumer trust and revisit intention at Beauty Care Clinic in Indonesia. *Archives of Business Research, 8*(6), 268–281. <https://doi.org/10.14738/abr.86.8489>.

Lashgari, M. H., Arefanian, S., Mohammadshahi, A., & Reza, A. (2015). Effects of the total quality management implication on patient satisfaction in the emergency Department of Military Hospitals. *3*(1), 2–6. <https://doi.org/10.581/jamm.26952>.

Liu, S., Li, G., Liu, N., & Hongwei, W. (2021). The impact of patient satisfaction on patient loyalty with the mediating effect of patient trust. *Inquiry (United States), 58*. <https://doi.org/10.1177/00469580211007221>.

Mohammed, S., Brahma, G. J., & Aderaw, S. (2020). Impact of Total Quality Management (TQM) on operational performance of Ethiopian pharmaceutical manufacturing plants. *SSRN Electronic Journal,* 4(August), 78–86. <https://doi.org/10.2139/ssrn.3479584>.

Mohd Isa, S., Lim, G. S. S., & Chin, P. N. (2019). Patients' intent to revisit with trust as the mediating role: Lessons from Penang Malaysia. *International Journal of Pharmaceutical and Healthcare Marketing, 13*(2), 140–159. <https://doi.org/10.1108/IJPHM-10-2017-0056>.

Nguyen, H., Calantone, R., & Krishnan, R. (2020). Influence of social media emotional word of mouth on institutional investors' decisions and firm value. *Management Science, 66*(2), 887–910.

Nguyen, N. X., Tran, K., and Nguyen, T. A. (2021). Impact of service quality on inpatients' satisfaction, perceived value, and customer loyalty: A mixed-methods study from a developing country. *Patient Preference and Adherence, 15*(August), 2523–2538. <https://doi.org/10.2147/PPA.S333586>.

Nguyen, T. L. H., & Nagase, K. (2019). The influence of total quality management on customer satisfaction. *International Journal of Healthcare Management, 12*(4), 277–285. <https://doi.org/10.1080/20479700.2019.1647378>.

Oliver, R. L., Rust, R. T., & Varki, S. (1997). Customer delight: Foundations, findings, and managerial insight. *Journal of Retailing, 73*(3), 311–336. <https://doi.org/10.1016/S0022-4359(97)90021-X>.

Qin, H., & Prybutok, V. R. (2013). A quantitative model for patient behavioral decisions in the urgent care industry. *Socio-Economic Planning Sciences, 47*(1), 50–64. <https://doi.org/10.1016/j.seps.2012.08.003>.

Robinson, S., Orsingher, C., Alkire, L., De Keyser, A., Giebelhausen, M., Papamichail, K. N., … Temerak, M. S. (2020). Frontline encounters of the AI kind: An evolved service encounter framework. *Journal of Business Research, 116*, 366–376.

Sabella, A., Kashou, R., & Omran, O. (2014). Quality management practices and their relationship to organizational performance. *International Journal of Operations and Production Management, 34*(12), 1487–1505. <https://doi.org/10.1108/IJOPM-04-2013-0210>.

Shabbir, S., Kaufmann, H. R., & Shehzad, M. (2010). Service quality, word of mouth and trust: Drivers to achieve patient satisfaction. *Scientific Research and Essays, 5*(17), 2457–2462.

Sholeh, M., & Chalidyanto, D. (2021). The effect of service quality on loyalty through patient satisfaction in outpatient of hospital X, Malang. *JMMR (Jurnal Medicoeticolegal Dan Manajemen Rumah Sakit), 10*(2), 148–157. <https://doi.org/10.18196/jmmr.v10i2.10239>.

Siripipatthanakul, S. (2021). Service quality, patient satisfaction, word-of-mouth, and revisit intention in a Dental Clinic, Thailand. *International Journal of Trend in Scientific Research and Development,* 5(5), 832–841.

Sweeney, J. C., Soutar, G. N., & Mazzarol, T. (2008). Factors influencing word of mouth effectiveness: Receiver perspectives. *European Journal of Marketing, 42*(3–4): 344–364. <https://doi.org/10.1108/03090560810852977>.

Yilma, T. A., Gebretekle, G. B., & Fenta, T. G. (2021). Patient satisfaction with HIV/AIDS services in health centers of East Shoa Zone, Oromia, Ethiopia: A cross-sectional study. *Health Services Insights, 14*. <https://doi.org/10.1177/11786329211003106>.

Zhou, W. J., Wan, Q. Q., Liu, C. Y., Feng, X. L., & Shang, S. M. (2017). Determinants of patient loyalty to healthcare providers: An integrative review. *International Journal for Quality in Health Care, 29*(4), 442–449. <https://doi.org/10.1093/intqhc/mzx058>.

SAMUEL-SOMA M. AJIBADE, OLUWADARE JOSHUA OYEBODE, SUSHOVAN CHAUDHURY AND ANTHONIA OLUWATOSIN ADEDIRAN

# 3 A Scientometric Analysis of Swarm-based Metaheuristics Approach

ABSTRACT
Particle-swarm Optimisation Techniques (POT) is a swarm-based metaheuristic approach that uses candidate solutions, or particles, depending on selected quality indicators to optimize computational problems. There has been an overwhelming amount of research published, but not nearly as many that critically evaluate the influence of this research on recent scientific advances. This prompted a look into the POT research publishing environment and trends. Scopus publications published between 2001 and 2021 were found and analyzed with scientometric qualitative analysis and the Preferred Reporting Items for Systematic Reviews and Meta-Analyses (PRISMA) guidelines. POT's growing popularity as a tool for solving computational problems is reflected in a 21,362.50 % increase in the number of papers devoted to the topic. The research landscape on POT is dominated by a small number of highly productive authors, institutions, and countries. The term "particle-swarm optimisation (PSO)" was found to be used more frequently than any other term in a co-occurrence analysis. Findings from this work may help academics and policymakers better understand the promise and challenges of POT research in light of other, similar algorithms in the literature.

Keywords: Swarm-based approach, metaheuristics technique, particle swarm, optimization techniques, scientometric analysis

## Introduction

Particle-swarm optimisation (PSO) is a metaheuristic approach to optimization and problem solving in computing which includes continually enhancing one or more candidate solutions depending on the specified quality measures. PSO is a population-based method for finding the optimal solution to a computational problem (Alsharif et al., 2015; Eberhart & Kennedy, 1995). The original PSO algorithm was developed by Eberhart and Kennedy (1995) to optimize swarm behavior in ways similar to those of birds and schooling fish. The algorithm uses a cyclical approach to find a solution. Particles representing potential solutions are shifted across the search space at a predetermined place and speed using a numerical equation (Kennedy, 1997). Best-case scenario, the particle's path across the search space is governed by its best-known local position (Malczewski, 2018). Each particle in Yang et al. (2015) uses its own intelligence as well as the knowledge of the group to guide its behavior. Particle swarm optimisation (PSO) is supposedly stimulated by utilizing groups of randomly positioned particles whose positions and velocities are determined by the N decision parameters of the optimization problem (Alsharif et al., 2015).

## Literature Review

Since its inception, the PSO algorithm has been used to address nonlinear issues across a wide range of disciplines, including economics, ecology, and technology. This is mostly because the technique has many advantages over other optimization algorithms (Alsharif et al., 2015), such as lower computing costs, better performance, and fewer changeable parameters. For numerical problem solving and optimization, the method is regarded as a fast, reliable, and straightforward procedure (Van den Bergh & Engelbrecht, 2006). Since POT's swarm intelligence approach is built on heuristics and does not suffer from slow convergence or dimensionality difficulties, it is a more efficient alternative to analytic approaches, especially for use in large-scale settings (Del Valle et al., 2008; Ajibade, 2020). POT also has the added benefit of not needing evolution operators like crossover and mutation (Wang et al., 2010). Ma et al. (2011) concluded that PSO's wide range of use is due to the fact that it is both versatile and easy to use, allowing it to carry out operations with few parameters.

In addition, POT is widely used for predicting a wide variety of established systems. Economic dispatch (ED) in power systems is studied in Rabbi et al. (2022), which used the POT's real-world generator restrictions as a research tool. To solve numerical benchmark problems, Vesterstrøm and Thomsen (2004) shown the potential of combining POT with evolutionary algorithms (EAs). POT, as demonstrated by Del Valle et al. (2008), can be used to solve non-linear optimization issues in many contexts involving power systems.

To show POT's applicability to fundamental investigations in chemistry and materials science, Wang et al. (2010) looked into its application to crystal structure prediction. In the PSO research, crystal structures were determined using the combined application of global reduction of free-energy surfaces and total-energy computations. As a result of reducing the number of variables in the optimization and search spaces and speeding up the convergence of the structure, the findings proved the technique's computational efficiency as well as low cost. Ma et al. (2011) and Masoomi et al. (2013) analyzed and solved land-use allocation issues using a GIS and the PSO algorithm (POT). Scientists have realized that GIS could benefit from integrating swarm intelligence-based optimization strategies. The strategy relies on the concept of a loose confederation of self-reliant individuals cooperating to address a challenge.

The study's overarching goal is to provide an in-depth analysis of the current state of Particle-swarm Optimization Algorithms in the fields of academia and industry. It is hoped that academics in academia, industry, and policymaking would be able to use the study's findings to better understand the POT and its advantages and disadvantages in comparison to alternative algorithms.

## Methodology

In this study, the publication trends and bibliometric analysis of published studies on the PSO were critically analyzed. The PRISMA (Preferred Reporting Items for Systematic Reviews and Meta-Analyses) technique was used to select, extract, and evaluate the published documents on the keywords from the Elsevier Scopus database. The selection phase involved designing and executing a search string based on the TITLE-ABS-KEY ("particle swarm" AND "optimi*ation" OR "algorithm") criteria. Due to the cumbersome and complexity of the large data set returned, the search criteria were revised to screen out unwanted publications. Hence, the search criterion: TITLE ("particle swarm" AND "optimi*ation" OR "algorithm") was executed in the Scopus database to obtain only documents with the keywords contained in the title of each recovered publication. The search was screened by limiting the document types to only Articles, Conference Papers, and Reviews as well as language to English and publication year from 2001 to 2021. The final selection of were subsequently analyzed to examine the publication trends. Next, bibliometric analysis was performed using the VOSviewer to deduce the network of co-authors, keyword occurrences, and citation trends on PSO. Lastly, a systematic literature review of the recent scientific and technological advancements on the topic was performed based on the selection of the benchmark publications (defined as published documents cited 50 or more times) on POT.

The publication patterns, significant stakeholders, sponsors, and nations actively participating in POT research in the literature were analyzed and calculated using Microsoft Excel (version 2016) once we retrieved the relevant documents. To further investigate the co-authorship, keyword occurrence, and citation tendencies in POT study, a bibliometric analysis was conducted using VOSviewer (Version 1.16.17). Finally, a PRISMA-guided systematic literature review was conducted to assess the field's scientific progress over the study period.

## Results and Discussion

### *General Publication Trends*

From 2001 through 2021, Figure 3.1 shows the overall trends in POT research publications as inferred by the Scopus database. The number of scholarly articles covering this subject increased steadily from 2001's low of 8 to 2021's high of 1,717 (a gain of 21,362.50 %), with 2019 seeing the greatest annual total of 1,802.

In total, 22,974 articles were uncovered from the time period under consideration. Between 2005 and 2010, the number of publications climbed from 304 to 1,517, a 399.013 % rise. From the looks of things, it's safe to assume that POT studies have attracted a lot of attention from experts in the field. The increasing popularity of using stochastic methods like PSO to solve optimization problems may be a result of more research into the topic. Examples of PSO's use in the electricity sector include the reduction of fuel prices, power losses, voltage profiles, and the improvement of system stability (Ajibade et al., 2022). According to Figure 3.2, the most

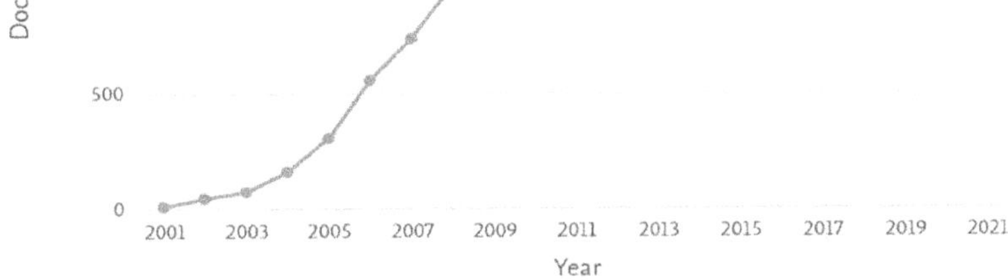


Figure 3.1: General publication trends on *POT* research (2001–2021)

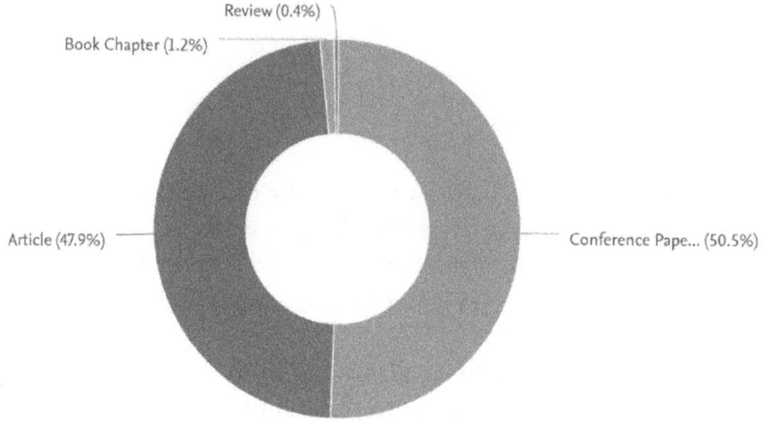


Figure 3.2: Document types for *POT* research (2001–2021)

common types of POT research papers are conference papers, articles, book chapters and review papers.

As has been noted, POT researchers do not follow the standard practice of publishing publications but instead publish their work in conference proceedings. This trend toward publishing in conference proceedings suggests a lively exchange of ideas and a fast pace of progress. The analysis also found that numerous publications and reviews on the subject had been published in the past.

Peer-reviewed reviews provide academics with in-depth data on the expansion and advancement of any given field's technology and science. The academic and scientific publishing communities, as well as those attempting to quantify the influence of research in any discipline, now place significant weight on the prestige and renown of journals. Figure 3.4 displays the most cited publications pertaining to POT studies.

## Major Research Stakeholders

Major research stakeholders can be analyzed to provide insight into the research landscape and scientific advancements in any discipline. All the

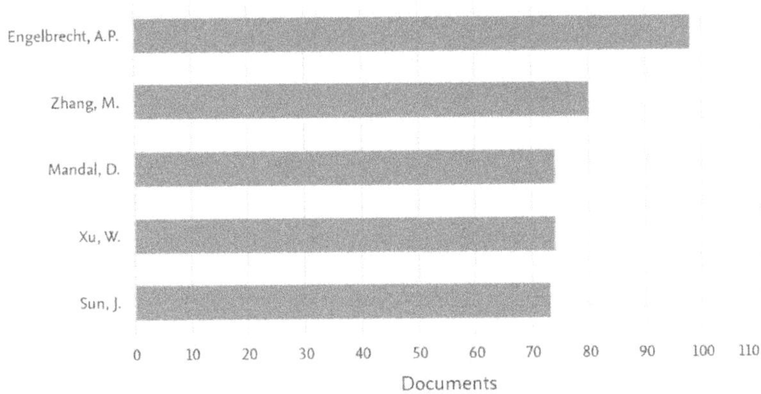


Figure 3.3: Top researchers on *POT* research (2001–2021)

people, institutions, groups, and nations that are actively engaged in a certain area of study. As a result, the leading scientists currently conducting research in the topic are depicted in Figure 3.3.

Andries P. Engelbrecht has done the most research on this issue, with 98 publications (0.43 % of the TPP). Mengjie Zhang, Durbadal Mandal, Wenbo Xu, and Jun Sun are also important researchers in this subject; they have (0.35 %), (0.32 %), and (0.32 %) published documents and (%TPP, respectively). As was subsequently discovered, leading researcher A.P. In addition to Engelbrecht's home base at South Africa's Stellenbosch University.

Over 200 articles have been produced by the top 5 affiliations on the issue during the study period. Similar to POT research, four of the top five universities doing this type of study are located in China. The University of Technology Malaysia is the only one of these that is not located in Malaysia. Figure 3.4 displays the next step, which was an analysis of the research activity in countries other than China and Malaysia.

According to the results, China, India, Iran, the United States, and Taiwan have the most researchers and associations working on POT-related topics.

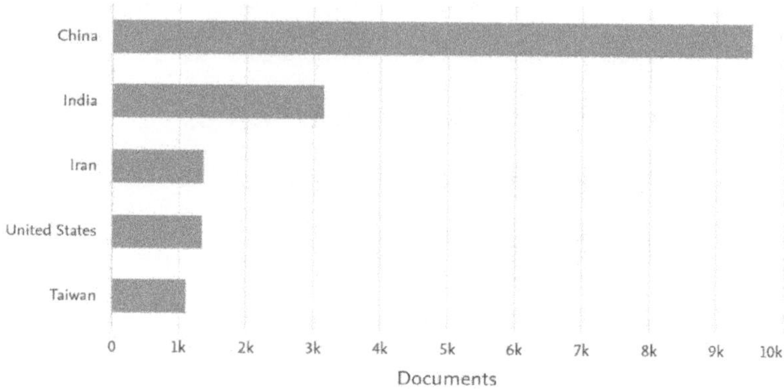


Figure 3.4: Top countries on *POT* research (2001–2021)

## Top Cited Publications

This study also analyzed the most-cited papers in the field of POT research in order to shed light on the significance of their contributions to the field. Based on the screening of articles with over 1,000 citations over the time period under consideration, Table 3.1 provides an overview of the top 10 most cited publications on the issue.

Based on the data collected, it is clear that the most-cited publications on this issue have received anywhere from 1,049 to 3,702 citations since the study's inception. Publication in the "Proceedings of the IEEE Conference on Evolutionary Computation" of "Innovations, software, and tools for particle-swarm optimisation" by Eberhart and Shi (2001) has received 3,702 citations, making it the most referenced paper ever. Table 3.1 illustrates that, on average, POT research studies receive many citations, which is indicative of their high level of research impact and academic interest. Because of this, it is expected that the volume of work done on POT subjects, the number of works citing those works, and the amount of money spent on these subjects will all rise in the coming years. New scholars, institutions, countries, and partnerships in this area will all help to propel this scenario forward. Within the POT Research, bibliometric

Table 3.1: Top 10 cited publications on POT

| References | Title | Journal/Source Title | Cited by |
|---|---|---|---|
| Eberhart and Shi (2001) | Innovations, software, and tools for particle-swarm optimization | Proceedings of the IEEE Conference on Evolutionary Computation, ICEC | 3,702 |
| Coello Coello et al. (2004) | Particle-swarm optimisation for goal-constrained multi-objective optimization | IEEE Transactions on Evolutionary Computation | 3,082 |
| Liang et al. (2006) | An all-inclusive multi-modal function optimizer based on a learning particle-swarm optimizer | IEEE Transactions on Evolutionary Computation | 2,818 |
| Trelea (2003) | Convergence analysis and parameter selection for the particle swarm optimisation algorithm | Information Processing Letters | 2,145 |
| Robinson and Rahmat-Samii (2004) | Particle-swarm optimization in electromagnetics | IEEE Transactions on Antennas and Propagation | 1,847 |
| del Valle et al. (2008) | Particle-swarm optimisation: fundamentals, variations, and power-system applications | IEEE Transactions on Evolutionary Computation | 1,726 |
| Ajibade et al. (2020) | Particle-swarm optimisation is used to figure out the structure of crystals. | Physical Review B—Condensed Matter and Materials Physics | 1,599 |
| Coello Coello and Lechuga (2002) | Multiple Objective Particle-Swarm Optimisation (MOPSO): A Proposed Approach | Proceedings of the 2002 Congress on Evolutionary Computation, CEC 2002 | 1,482 |
| Zhan et al. (2009) | Constraint-based particle-swarm optimization | IEEE Transactions on Systems, Man, and Cybernetics, Part B: Cybernetics | 1,441 |

analysis was also performed to further investigate the topic's current state and potential future outcomes.

## Bibliometric Analysis

Bibliometric analysis is a cutting-edge method for investigating the current state of knowledge and future prospects in any area of study (Trelea, 2003; del Valle et al., 2008). It uses mathematical/statistical algorithms to locate relevant research publications or published documents and then screen them for co-authorships, keyword occurrences, and (co-citations) on a specific topic (Ajibade et al., 2020; Coello Coello & Lechuga, 2002).

## Co-author Analysis

This is illustrated by a network diagram depicting POT Research co-authorship (Figure 3.5). There were at least 10 papers used for the analysis,

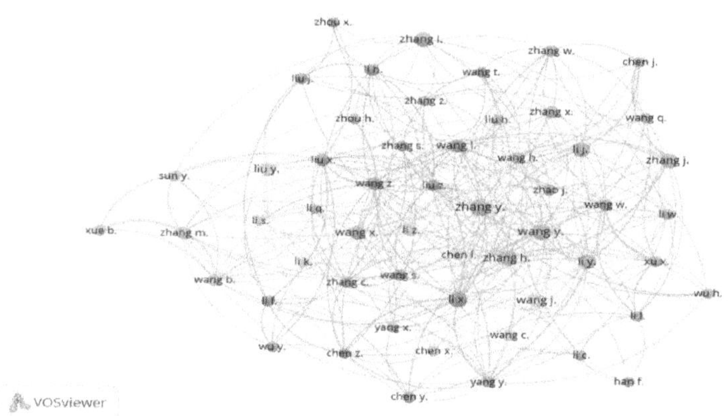


Figure 3.5: Network visualization for co-authorship on *POT*

and each author was cited at least five times. The search yielded a total of 4,781 authors, 56 of which met the criteria and were thus chosen for the co-analysis. The red cluster, which is the most prominent, has 15 authors, including some of the most well-known names in the field.

## Keyword Co-occurrence

Research landscape in any topic can be analyzed through the literature by tracking how often certain keywords appear (Zhan et al., 2009; Sun et al., 2004). As a result, this study analyzed the frequency with which certain keywords appeared in the POT literature. Figure 3.6 is a network visualization of the topic's most frequently used terms.

According to the study's foundational minimum keyword frequency of 25, a total of 14,347 keywords were identified. The results of the search yielded 126 keywords and 6 clusters, each containing 7–36 items. Particle-swarm optimisation (PSO), "particle-swarm optimisation," and "optimisation" were found to be the most common search terms, appearing 1726 times, 848 times, and 612 times, respectively, with a total link strength of

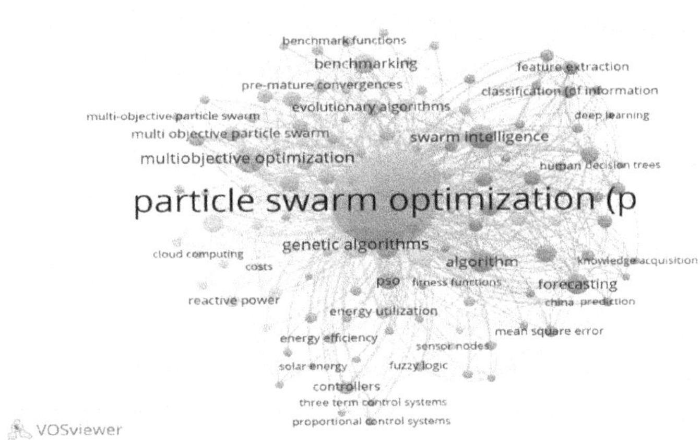


Figure 3.6: Network visualization of co-occurring keywords on *POT*

7,988, 4,000, and 3,404, respectively. The study's title and search query played a role in attracting a large number of relevant results, so this isn't entirely coincidental. Multi-objective optimization, genetic algorithms, swarm intelligence, benchmarking, and forecasting are some prominent terms that emerged from the data analysis. The most common search terms on POT studies reveal connections and overlaps between fields including IT, AI, math, and stats. As a result, it's safe to assume that the issue encompasses many different areas of study.

## Citation Analysis

Publications' research and scientific significance can be gauged by the number of times they are cited by other studies. Accordingly, a citation analysis of POT studies was conducted to look at how often particular authors, publications, and nations were cited in the study's body of work. For the author-based citation analysis, we used the following search parameters: a minimum of 10 documents and a maximum of 5 citations per author. According to the findings, 4,781 authors met the requirements, but only 56 were chosen for in-depth examination. The author-based citations in POT research are visualized as a network (Figure 3.7) below.

According to the results, a group of 55 authors with the most ties together can be broken down into 6 distinct clusters of anywhere from 4 to 15 writers. The most significant author is the red cluster, which includes such names as Zhang L., Wang X, and Li W. In terms of total connection strength and co-citations, the cobalt blue cluster's Zhang Y. stands out as the leader. Overall, the results show that the most authoritative voices in the field regularly refer to and cite published materials on the subject. Furthermore, the high number of citations and co-citations shows that researchers in the subject engage in extensive cross-institutional and international research. Then, the most-cited papers in the field were analyzed for their citation patterns. Using the network visualization map presented in Figure 3.8, this analysis seeks to understand the subject's benchmark studies. There are clearly 16 distinct groupings of data, each containing

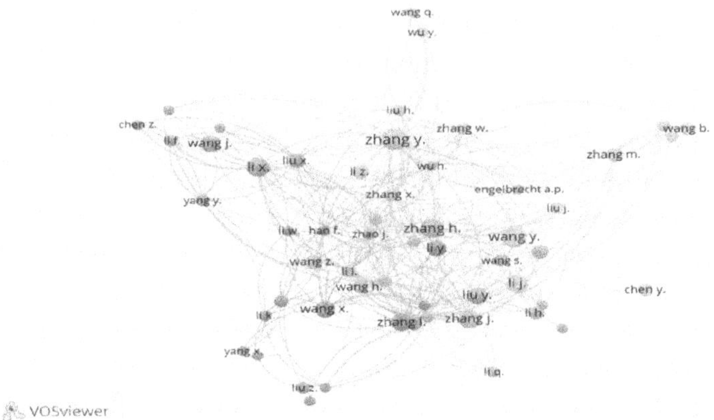


Figure 3.7: Network visualization for the author-based citations on *POT*

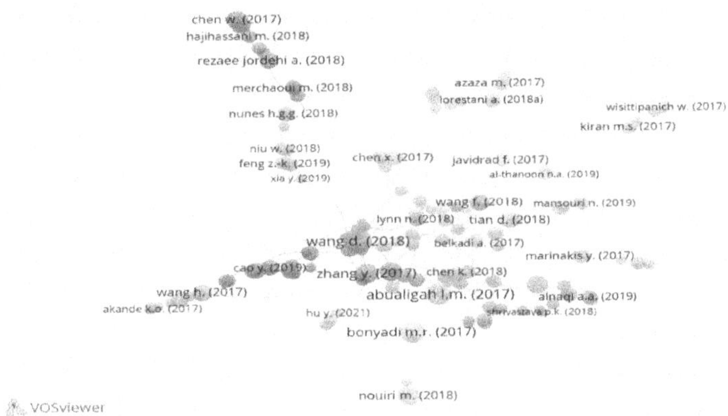


Figure 3.8: Network visualization map of the co-citations among top published/cited documents on *POT* research

between three and eight elements. The works of Cao Y. (2019) make up the largest and most prominent red cluster, which has the strongest links of any cluster, indicating a significant number of co-citations and, thus, an impactful body of work.

## Conclusions

This research provided a thorough assessment of the state of study and recent scientific advances in the field of particle-swarm optimization algorithms (POT). Using VOSViewer, the researchers were able to conduct bibliometric analysis of all the articles about the topic that were included in the Scopus database Survey on (2001–2021) to achieve their research aims. Over the course of the study's 20-year time frame, researchers looked at how POT had changed in terms of citations, major players in the field, funding sources, and publications. The results showed that between 2001 and 2021, the number of papers devoted to POT increased by a factor of 21,362.50 %, from 8 to 1,717. This dramatic rise can be related to the expanding use of PSO to address optimization challenges in research, business, and government. According to the bibliometric analysis conducted on this topic, there are numerous publications, partnerships, and citations in this area across all affiliations/organizations and countries. Therefore, it's reasonable to conclude that there are broad research and scientific themes at play here that cut across disciplines.

## Editors Note

The examination of a scientometric analysis of swarm-based metaheuristics approaches holds significant importance for sustainability across various domains. Swarm-based metaheuristics, such as particle swarm optimization and ant colony optimization, offer innovative solutions to complex optimization problems by simulating collective behaviors inspired by natural systems. By conducting a scientometric analysis of research output in this area, researchers can identify trends, gaps, and emerging topics, thus informing future research directions, and facilitating knowledge dissemination. Furthermore, the application of swarm-based metaheuristics has the potential to address sustainability challenges across diverse sectors, including energy, transportation, manufacturing, and environmental management. By optimizing resource allocation, minimizing energy consumption, and improving operational efficiency, these approaches contribute to the sustainable use of resources and the mitigation of environmental impacts. Therefore, a scientometric analysis of swarm-based metaheuristics not only enhances our understanding of optimization techniques but also provides valuable insights for advancing sustainability goals in various field.

## References

Ajibade, S. (2020). Particle-swarm optimization with chaotic dynamic weight for feature selection enhancement. *Journal of Science, Engineering Technology and Management, 1*(2), 1–5.

Ajibade, S. S. M., Ahmad, N. B., & Shamsuddin, S. M. (2020). A data mining approach to predict academic performance of students using ensemble techniques. In *Intelligent systems design and applications: 18th international conference on Intelligent Systems Design and Applications (ISDA 2018) held in Vellore, India, December 6–8, 2018, Volume 1* (pp. 749–760). Springer International Publishing.

Ajibade, S. S. M., Ahmad, N. B. B., & Zainal, A. (2020, July). A hybrid chaotic particle-swarm optimization with differential evolution for feature selection. In *2020 IEEE Symposium on Industrial Electronics & Applications (ISIEA)* (pp. 1–6). IEEE.

Ajibade, S. S. M., Zaidi, A., Tapales, C. P., Ngo-Hoang, D. L., Ayaz, M., Dayupay, J. P., … Adediran, A. O. (2022, December). Data mining analysis of online drug reviews. In *2022 IEEE 10th Conference on Systems, Process & Control (ICSPC)* (pp. 247–251). IEEE.

Alsharif, M. H., Nordin, R., & Ismail, M. (2015). Exploiting coexistence between UMTS and LTE for Greener cellular networks with particle-swarm optimization. *Wireless Personal Communications, 85*(3), 623–639.

Coello Coello, C. A., & Lechuga, M. S. (2002). MOPSO: A proposal for multiple objective particle-swarm optimization. In *2002 Congress on Evolutionary Computation, CEC 2002*. Honolulu, HI: IEEE Computer Society.

Coello Coello, C. A., Pulido, G. T., & Lechuga, M. S. (2004). Handling multiple objectives with particle-swarm optimization. *IEEE Transactions on Evolutionary Computation, 8*(3), 256–279.

Del Valle, Y., Venayagamoorthy, G. K., Mohagheghi, S., Hernandez, J.-C., & Harley, R. G. (2008). Particle-swarm optimization: Basic concepts, variants and applications in power systems. *IEEE Transactions on Evolutionary Computation, 12*(2), 171–195.

Eberhart, R., & Kennedy, J. (1995). Particle-swarm optimization. In *Proceedings of the IEEE international conference on neural networks*. Citeseer.

Eberhart, R. C., & Shi, Y. (2001). Particle-swarm optimization: Developments, applications and resources. In *Congress on Evolutionary Computation 2001*. Soul.

Kennedy, J. (1997). The particle swarm: Social adaptation of knowledge. In *Proceedings of 1997 IEEE International Conference on Evolutionary Computation (ICEC'97)*. IEEE.

Liang, J. J., Qin, A. K., Suganthan, P. N., & Baskar, S. (2006). Comprehensive learning particle-swarm optimizer for global optimization of multimodal functions. *IEEE Transactions on Evolutionary Computation, 10*(3), 281–295.

Ma, S., He, J., Liu, F., & Yu, Y. (2011). Land-use spatial optimization based on PSO algorithm. *Geo-spatial Information Science, 14*(1), 54–61.

Malczewski, J. (2018). 1.15—Multicriteria analysis. In B. Huang (Ed.), *Comprehensive geographic information systems* (pp. 197–217). Oxford: Elsevier..

Masoomi, Z., Mesgari, M. S., & Hamrah, M. (2013). Allocation of urban land uses by multi-objective particle-swarm optimization algorithm. *International Journal of Geographical Information Science, 27*(3), 542–566.

Rabbi, F., Ayaz, M., Dayupay, J. P., Oyebode, O. J., Gido, N. G., Adhikari, N., … Bassey, M. A. (2022, July). Gaussian map to improve firefly algorithm performance. In *2022 IEEE 13th Control and System Graduate Research Colloquium (ICSGRC)* (pp. 88–92). IEEE.

Robinson, J., & Rahmat-Samii, Y. (2004). Particle-swarm optimization in electromagnetics. *IEEE Transactions on Antennas and Propagation, 52*(2), 397–407.

Sun, J., Feng, B., & Xu, W. (2004). Particle-swarm optimization with particles having quantum behavior. In *Proceedings of the 2004 Congress on Evolutionary Computation, CEC2004*. Portland, OR.

Trelea, I. C. (2003). The particle-swarm optimization algorithm: Convergence analysis and parameter selection. *Information Processing Letters, 85*(6), 317–325.

Van den Bergh, F., & Engelbrecht, A. P. (2006). A study of particle-swarm optimization particle trajectories. *Information sciences, 176*(8), 937–971.

Vesterstrøm, J., & Thomsen, R. (2004). A comparative study of differential evolution, particle-swarm optimization, and evolutionary algorithms on numerical benchmark problems. In *Proceedings of the 2004 Congress on Evolutionary Computation, CEC2004*. Portland, OR.

Wang, Y., Lv, J., Zhu, L., & Ma, Y. (2010). Crystal structure prediction via particle-swarm optimization. *Physical Review B, 82*(9), 094116.

Yang, X.-S., Chien, S. F., & Ting, T. O. (2015). *Bio-inspired computation in telecommunications.* Elsevier. <https://doi.org/10.1016/C2014-0-00501-1>

Zhan, Z. H., Zhang, J., Li, Y., & Chung, H. S. H. (2009). Adaptive particle-swarm optimization. *IEEE Transactions on Systems, Man, and Cybernetics, Part B: Cybernetics, 39*(6), 1362–1381.

DR. JAPHETH NUHU AHMED, FIGEN YEŞILADA, HAYFORD ASARE OBENG, DAGNU HAILE TESSEMA, DOMINIC J. KAMARA, DENNIS AFFUM OSEI AND PETER C. MORNYEY

# 4 The Mediating Effect of Patient Trust on the Link between Relationship Marketing and Hospital Brand Image in Public Hospitals-Nigeria

ABSTRACT

Background: Healthcare institutions are currently competing more fiercely than ever, although in public sector the competition seems less intense. Health institutions must maintain a relationship with the patients who use their services in order to thrive in this context of increased competition.

Aim/Purpose: The aim of this study is to examine the mediating effect of patient trust on the link between relationship marketing and hospital brand image in selected public hospitals.

Approach: The measurement instruments employed in the research were relationship marketing, patient trust, and hospital brand image. The population under investigation consisted of outpatients who received medical attention from selected public hospitals operating in Gombe State, Nigeria. The study was carried out between January 1 and February 28, 2023. Principally utilizing the convenient sampling method, the study was executed on a total of 350 patients who sought treatment from selected public hospitals. Data analysis was done using Amos Version 21 software.

Results: According to the research's findings, relationship marketing was found to positively impact the hospital's brand image. Furthermore, relationship marketing had a direct and positive influence on patient trust. Notably, patient trust was identified as having a significant and affirmative influence on the hospital's brand image. The influence of hospital relationship marketing on hospital brand image was further enhanced by the effect of patient trust. In other words, patient trust plays a mediating role in how relationship marketing affects the perception of the hospital's image.

Limitation: One significant limitation of the study is that it only uses quantitative data. Additionally, there are other limitations based on the geography, industry, and sample size. As a result, it is difficult to generalize the findings across various businesses and countries.

Implication: Based on the study's results, the marketing and management departments of hospitals should focus on the patients' perspectives regarding relationship marketing and trust to establish a favorable hospital image.

Keywords: Brand image, hospitals, trust, relationship marketing

## Introduction

There is increased competition among medical institutions in the contemporary environment. To survive, health facilities must create long-lasting relationships with their patients. Relationship marketing is recognized as a highly effective strategy that is essential in creating long-lasting connections with patients (Suryadi, Pasinringi, & Kadir, 2022). Building sustainable relationships that are characterized by loyalty, satisfaction, and trust is one of the main goals of relationship marketing (Solangi et al., 2019).

Hospital relationship marketing significantly influences a hospital's brand image. A hospital's reputation can be greatly impacted by how it communicates and engages with its patients, their families, and the general public (Irizarry, DeVito, Dabbs, & Curran, 2015). A strong brand reputation can help a hospital stand out in the cutthroat healthcare market of today and draw in and keep patients.

Building lasting, mutually beneficial relationships with patients and their families is the goal of relationship marketing (Fazal-e-Hasan et al., 2020). Understanding their preferences and wants is necessary to customize services to satisfy those needs. Hospitals can achieve this by providing a satisfying patient experience, which develops loyalty and promotes referrals from friends and family.

A hospital's emphasis on relationship marketing conveys that it cares about its patients in more ways than one. As a result, people may have positive opinions of the hospital's brand and its services. Patients are more likely to talk favorably about their experiences and suggest the hospital to others if they feel appreciated and well-cared for (Geese, Willener & Spichiger, 2021).

Relationship marketing can also assist hospitals in setting themselves apart from their rivals. Hospitals can differentiate themselves in a crowded market by providing individualized services and going above and beyond to fulfill patients' demands (Allahham, Mouselli & Jakovljevic, 2022). This can support the development of a powerful and distinctive brand identity that connects with customers and the local community.

Marketing to hospitals is an essential part of creating a strong brand. Hospitals may develop a strong brand that encourages loyalty, trust, and favorable word-of-mouth recommendations by putting a priority on patient engagement and satisfaction (Limbu et al., 2020).

## Originality and Significance of the Study

This article's goal is to explore the role of patient trust as a mediator in the connection between relationship marketing and hospital brand image in public healthcare institutions. Although there is a sizable amount of research on relationship marketing, very few studies have looked at how it relates to customer loyalty, repurchase intention, and word-of-mouth advertising. To the best of the authors' knowledge, no earlier studies in the healthcare sector have looked at the impact of patient trust on the connection between relationship marketing and hospital brand image. It is imperative to note that relationship marketing studies has mostly been done in industries other than healthcare. Our research seeks to fill this knowledge gap and contribute to the existing literature. The study highlights the consequence of building enduring relationships with patients and establishing their trust, as it can critically add to enhancing the perception of a hospital's brand.

## Literature Review

### *Relationship Marketing*

According to Zewdu (2022), a multiservice company uses relationship marketing as a strategy to draw customers in, keep them, and strengthen their bonds with them. Berne-Manero and Marzo-Navarro (2020) also emphasized that the primary objective of all relationship marketing activities is to establish, cultivate, and sustain effective business relationships. Relationship marketing, as defined by Wolter et al. (2022), entails ongoing collaboration between the actor and the target market in order to draw in, grow, and retain customers. Relationship marketing, according to Veselova (2022), is a strategy used by companies to build enduring, mutually beneficial relationships with their clients through cooperative, voluntary contacts.

### *Hospital Brand Image*

Brand image is the receiver's view of a company's identity and their interpretation of its attributes from many sources (Boafo et al., 2020). This complicated perception frequently influences how people see the organization. According to studies by Darmawan and Grenier (2021), a company's image can be assessed based on elements including ease of access to services, the quality of those services, one-on-one interactions, and security reputation. Contrarily, corporate image relates to how an organization presents itself to its audiences as well as how those audiences see and understand that presentation (Febriyantoro, 2020).

### *Patient Trust*

Consumer trust is the hope that service providers will keep their word and be dependable (Setiawan et al., 2020). According to Chen et al.

(2015), trust is the readiness to rely on a brand, product, or service based on its perceived expertise, benefit, and trustworthiness (Konuk, 2021). Building patient trust in the healthcare sector can be done by providing individualized care, effective communication, and high-quality services (Du et al., 2020). The demands and preferences of patients are taken into account when providing personalized care. Clear and compassionate communication with patients is necessary, as is listening to their concerns and providing fast responses to their inquiries (Colgan et al., 2021). Because patients who receive high-quality care are more likely to trust their healthcare providers and promote them to others, high-quality services are crucial for fostering patient trust (Biddle et al., 2022).

## Hypotheses Development

### *The Impact of Relationship Marketing on Hospital Brand Image*

According to previous research, when businesses use relationship marketing, it may help consumers form favorable opinions of the company (Ngoma & Ntale, 2019). According to Hayati et al. (2020), businesses take steps to create and sustain positive customer relationships to draw in clients and improve their brand image. A strong corporate image can be created, and customer loyalty can be increased, by businesses taking steps to establish long-term relationships with their clients, such as fostering positive interactions between staff and clients (Özkan et al., 2020).

> ***H1:*** *Relationship marketing has a positive and significant effect on Hospital brand image*

## The Impact of Hospital Relationship Marketing on Patient Trust

Healthcare professionals can gain the trust of their patients by using a variety of strategies. Personalized care, which involves tailoring the treatment given to match the specific requirements and preferences of patients, is one efficient strategy (Stumpp & Sauer-Zavala, 2022). Effective communication, which includes direct and sympathetic interaction, attentive listening to patients' challenges, and rapid responses to their inquiries, is another key strategy (Ekman et al., 2022). Providing high-quality services is crucial for creating patient trust since people who receive excellent care are more inclined to believe in their healthcare providers and refer them to others (Lee, 2021).

According to the study by Sousa and Alves (2019), relationship marketing's essential elements of individualized care and clear communication have a beneficial effect on patients' trust and behavioral intentions. Similarly, Elizar, Indrawati, and Syah (2020) discovered that developing and sustaining strong relationships with patients depends on trust-building behaviors like empathy and good communication. The success of businesses in conducting business is greatly influenced by consumer trust, in this case, represented by patients (Tran & Vu, 2019). Therefore, it is crucial that service providers pay attention to patient trust. According to Abid (2023), relationship marketing is crucial for sustaining and cultivating strong ties with customers and forging long-lasting bonds and partnerships by earning their trust.

***H2:** Relationship marketing have a direct and significant impact on patient trust.*

## The Impact of Patient Trust on Hospital Brand Image

Intangible service transactions require brand trust, particularly in the healthcare sector (Cham, Lim, & Sigala, 2022). Because there are such high stakes in this industry, it is crucial to build and preserve client trust. The results of numerous studies on the subject indicate that trust is built when one side has faith in the dependability and morality of their trading partner (Ryan, 2020; Kim, Lee, & Hwang, 2020; Konuk, 2019). A number of variables, including consumer perceptions of the brand, its perceived quality, its prestige, and the customer-focused conduct of its staff, all have an impact on consumer trust in healthcare brands (Mashayekhi et al., 2022). Customers' affective reactions to brands, which may be favorable or unfavorable, are referred to as brand attitudes (Warren, Carter, & McGraw, 2019). To effectively brand a service, these attitudes must be controlled. The more a brand is trusted, the higher the perceived quality (Atulkar, 2020). Customers' perceptions of the brand's superiority or inferiority are also influenced by their associations with it (Lacap, Cham, & Lim, 2021). The more closely a customer identifies with a brand, the more favorably they will perceive the quality of its services (Nguyen, Quach, & Thaichon, 2022).

Brand image is also connected to brand trust. It is well established in research that customers who have a high level of trust in a brand are more likely to stick with it (Campbell et al., 2020). As a result, it is generally accepted that brand trust leads to a favorable perception of brand image (Bernarto et al., 2020). So, based on increased brand trust, which includes trust, dependency, and confidence in a hospital, we predicted that higher levels of brand loyalty would be the consequence (Kholis & Ratnawati, 2021). Therefore, the following hypothesis is put forward:

***H3:*** *Patient trust has a positive significant effect on hospital brand image*

## The Mediating Effect of Patient Trust between Relationship Marketing and Hospital Brand Image

Patient trust may have an impact on the characteristics and reputation of the healthcare service provider (Ponathi et al., 2020). Reputation is described as "public information encompassing the concept of trustworthiness" or "the merit of trust" by Malhotra (2019). When a hospital has a bad image, it puts its credibility in jeopardy and acts as a relationship-building catalyst (Liao et al., 2020). When hospitals establish trust with patients utilizing relationship marketing strategies, patients are more likely to perceive the hospital positively, which leads to a positive brand image (Cham et al., 2022). In contrast, when patients do not trust the hospital, relationship marketing strategies may not be effective in building a positive brand image (Mohd Isa et al., 2019). Therefore, the resulting hypothesis is proposed:

***H4:*** *Patient trust mediates the link between Relationship marketing and Hospital brand image*

## Research Methods

A survey questionnaire was utilized to conduct the study, and the selection criteria included outpatients aged 18 years or older who received treatment at the selected hospitals in Gombe State from January 5 to February 18, 2023, and gave their consent to participate. In Wolf et al.'s (2013) investigation, they researched the effects of sample size constraints for popular forms of structural equation modelling (SEM), as well as the effects of model type, number of elements, the intensity of indicator loadings, and regression routes on the outcomes. We concluded that a minimum sample size of 350 cases was necessary to accurately identify the components and indicators of our study model. To ensure content validity, the assessments used in this study were modified from other studies. With responses ranging from "strongly disagree" (1) to "strongly agree" (5), a five-point Likert scale was adopted to operationalize the measuring items. The five-item trust scale was adapted from Han and Hyun (2015) and Gambarov et al. (2017). The measurement tool for relationship marketing consists of six items and was adapted from Mohd Isa, Lim, and Chin (2019). The hospital brand image measuring items consist of three items and were adapted from Cham et al. (2016). The statistical software SPSS (version 25.0) was used to analyze the collected data. In addition, the Amos 21.0 SEM program's confirmatory factor analysis (CFA) was employed to examine the connections and correlations between the endogenous constructs of the model.

## Data Analysis Method

Sekaran and Bougie served as the foundation for the data analysis strategy used in this investigation (2016). The construct validity and research hypotheses were then put to the test using a measuring model and a structural model. Data entry and first analysis were done using IBM SPSS software (Version 25), and the measurement and structural model were tested using AMOS (Version 21). In this study, the validity of the indicators, convergent validity, and reliability were assessed using different methods. The value of the indicators, convergence, and reliability were tested using load factor analysis, AVE, CR, and Cronbach's alpha. While discriminant validity looks at how disconnected the measurements of the various qualities in the model are, convergent validity assesses the extent of confidence in the assessment of a trait by its elements. The bootstrapping method was used to test the accuracy of sample estimates, including bias, volatility, ranges of confidence, and incorrect predictions. The evaluation of the hypotheses was completed using structural equation modelling (SEM). The Amos 21.0 SEM program was used for confirmatory factor analysis (CFA) to establish the connections and correlations between the model's endogenous constructs. The statistical package SPSS (version 25.0) was used for data analysis.

## Results

### *Reliability and Validity*

The reliability analysis of the scales was evaluated using Cronbach's alpha value. Construct reliability takes into account how well a set of

Table 4.1: Reliability

| Variables | Scales | Cronbach's Alpha |
|---|---|---|
| Relationship marketing (RMRKT) | 6 | 0.880 |
| Patient trust (PTR) | 5 | 0.856 |
| Hospital brand image (HBIM) | 3 | 0.842 |

indicators that represent a particular construct are stable and consistent throughout. Cronbach's alpha for the relationship marketing (RM), patient trust (PTR), and hospital brand image (HBIM) constructs were 0.880, 0.856, and 0.842, respectively. Table 4.1 contains the reliability results of the constructs, namely relationship marketing (RM), patient trust (PTR), and hospital brand image (HBIM). All of the latent variables had Cronbach's alpha scores above 0.70, indicating that the findings were internally consistent.

## Confirmatory Factor Analysis

The CFA was used to carry out the SEM. CFA evaluated the construct and the accurate assignment of variables by observing the variables.

The Chi-square/degrees of freedom ratio, the comparative fit index (CFI), the goodness-of-fit index (GFI), the adjusted GFI (AGFI), the RMSEA (root mean square error of approximation), and the P-value were the indexes that were used to evaluate the suitability of the model in general. Modification indices (MI) and standardized residuals were assessed to enhance the model's fitness. It was necessary for the MI values to be under 20 %. For item pairings having standardized residuals between 2.5 and 4.0, the residuals were assessed. The standardized regression weights that were used to build the model have to be greater than 0.5. The bulk of the standardized coefficients in Table 4.2 were over 0.5 and fell within the range of 0.617–0.855. The CR varied from 0.893 to 0.909, while the AVE values for the components ranged from 0.789 to 0.861. These measures (CR > 0.7; AVE > 0.5) both exceeded the standards. The overall indices

Table 4.2: Confirmatory factor analysis

| Construct Measures | Standardized Coefficients | Average Variance Extracted (AVE) | Composite Reliability (CR) |
|---|---|---|---|
| RM6 <---RMRKT | 0.617 | 0.789 | 0.909 |
| RM5 <--- RMRKT | 0.800 | | |
| RM4 <--- RMRKT | 0.769 | | |
| RM3 <--- RMRKT | 0.782 | | |
| RM2 <--- RMRKT | 0.687 | | |
| RM1 <--- RMRKT | 0.818 | | |
| TR1 <--- PTR | 0.734 | | |
| TR2 <--- PTR | 0.663 | 0.790 | 0.893 |
| TR3 <--- PTR | 0.821 | | |
| TR4 <--- PTR | 0.720 | | |
| TR5 <--- PTR | 0.750 | | |
| HBI1 <--- HBIM | 0.855 | 0.861 | 0.896 |
| HBI2 <--- HBIM | 0.735 | | |
| HBI3 <--- HBIM | 0.814 | | |

Chi-square = 110.647, Degrees of freedom = 74, Chi-square (CMIN/Df) = 1.495, CFI = 0.983

demonstrate the model's applicability. Since all observable variables could be analyzed, the model was suitable. Each measure connected to the construct was statistically significant, Tables 4.2 and 4.3, and Figure 4.1, present the results of the SEM

## Goodness-of-Fit

Hair et al. (1998) found adequate model goodness-of-fit requirement index values. Table 4.3 contain the data for our model. The RMSEA score of 0.038, which is below the allowable limit, demonstrates that the empirical covariance matrix and model are fitting appropriately after a thorough

Table 4.3: Goodness of good-fit

| Model Fit | Index | Scores | Range | Decision |
|---|---|---|---|---|
| Absolute fit | Chi-square | 110.647 | | |
| | Df | 74 | | |
| | CMIN/Df | 1.495 | < 5 | Accepted |
| | RAMSEA | 0.038 | < 0.08 | Accepted |
| | GFI | 0.958 | >0.85 | Accepted |
| Incremental fit | NFI | 0.952 | >0.90 | Accepted |
| | TLI | 0.979 | | Accepted |
| | CFI | 0.983 | | Accepted |
| Parsimony-Adjusted Measures | PRATIO | 0.813 | | Accepted |
| | PNFI | 0.774 | | Accepted |
| | PCFI | 0.800 | | Accepted |
| | AGFI | 0.940 | >0.80 | Accepted |

analysis. The TLI, NFI, and CFI indices are 0.979, 0.952, and 0.983, respectively, while the CMIN/DF ratio is 1.495, which is indicative of this fitting as well, while the GFI and AGFI values were 0.958 and 0.940. Hence, the reported values suggested that the study model was suitable.

## Hypothesis Testing

Table 4.4: Direct effect

| Hypotheses | | Coefficient | SE | C.R | P-value | Result |
|---|---|---|---|---|---|---|
| **H1** | Hospital brand image ← Relationship marketing | 0.134 | 0.075 | 2.188 | 0.029 | Supported |
| **H2** | Patients trust ← Relationship marketing | 0.194 | 0.070 | 3.157 | 0.002 | Supported |
| **H3** | Hospital brand image ← Patients trust | 0.232 | 0.068 | 3.666 | 0.000 | Supported |

## Mediation Effects

Based on the analysis results, it was found that there is a direct impact ($\beta$ = 0.164) of relationship marketing on hospital brand image. Upon incorporating the mediating role of patient trust, a positive impact ($\beta$ = 0.055) was observed, indicating a partial mediation result. Overall, the model suggested a positive ($\beta$ = 0.220) and statistically significant ($p < 0.001$) effect of relationship marketing on hospital brand image. This highlights the significance of relationship marketing in influencing hospital brand image, with its impact being further influenced by patient trust. The results support the hypothesis (H4), as presented in Tables 4.5–4.6 and Figure 4.1.

Table 4.5: Hypothesis testing result for mediation

| **Hypothesis** | **Path** | **Effect** | **P-Value** | **Result** |
|---|---|---|---|---|
| **H4** | HBIM ← PTR← RMRKT | 0.055 | 0.006 | Supported |

Table 4.6: Direct, indirect and total effect

| | | **RMAKETIN** | **PATRUST** | **HOSPIIMAGE** |
|---|---|---|---|---|
| **Direct Effects** | PATRUST | 0.221*** | 0.000 | 0.000 |
| | HOSPIIMAGE | 0.164*** | 0.250*** | 0.000 |
| **Indirect Effects** | PATRUST | 0.000 | 0.000 | 0.000 |
| | HOSPIIMAGE | 0.055*** | 0.000 | 0.000 |
| **Total Effects** | PATRUST | 0.220 | 0.000 | 0.000 |
| | HOSPIIMAGE | 0.220 | 0.232 | 0.000 |

The direct effect of Relationship marketing on patient trust and the indirect effect (mediated) of Relationship marketing on Hospital Brand Image are significantly different from zero at the 0.001 level ($p = 0.005$ two-tailed). This is a bootstrap approximation obtained by constructing two-sided bias- corrected confidence intervals.

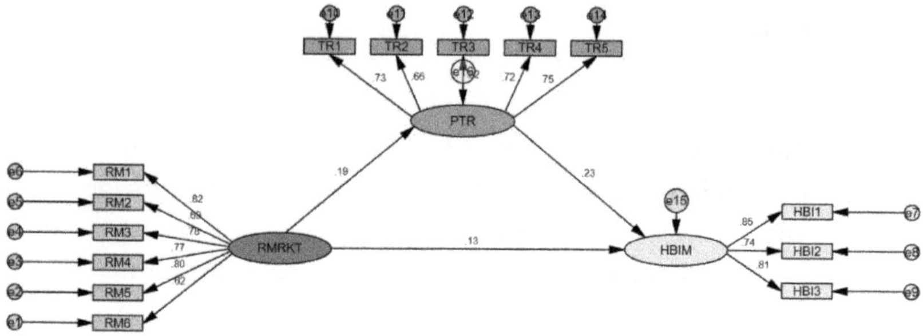


Figure 4.1: Structural model

## Discussion

The study began by examining the effect of relationship marketing on hospital brand image, followed by exploring its impact on trust and the role of trust in fostering a positive hospital brand image. Additionally, the mediating function of trust in the connection between relationship marketing and hospital brand image was examined.

### *H1: Relationship Marketing Has a Positive and Significant Effect on Hospital Brand Image*

The standardized coefficient revealed a significant positive impact ($\beta = 0.164^{***}$) of relationship marketing (RMRKT) on hospital brand image (HBIM). The reputation of the hospital as seen by the patients will therefore improve as a result of engaging in hospital relationship marketing. This result is consistent with Ashton, Scott, Solnet, and Breakey's (2010) study, which found that companies engage in customer relationship activities to attract customers and improve their brand image. Additionally, the actions taken by companies to sustain lasting relationships with clients, for example, friendly exchanges between staff and clients, contribute to the development of a positive corporate image and promote customer loyalty. Also, the results are in line with those of Chen and Chen (2014), who discovered that relationship marketing significantly improves hospital image.

### *H2: Relationship Marketing Has a Direct and Significant Impact on Patient Trust*

The standardized estimate of relationship marketing's impact on hospitals' brand image (TR <--- RMRKT) was 0.221***. This explains that relationship marketing (RMRKT) has a significant positive impact on trust (TR). In the context of healthcare, relationship marketing focuses on meeting patient requirements, wants, and desires through active listening, quick responses, and flexible provision. Hospitals can gain insight into patients' thoughts by carefully examining and actively monitoring the services they seek out concerning their medical needs. According to the current study's findings, relationship marketing significantly and directly affects patient trust. This outcome is in line with earlier research, which discovered that relationship marketing can increase consumer trust (Heffernan et al., 2008). As a result, hospitals can develop good relationships with their patients by promoting open communication. This will enable the hospitals to better understand how the patients feel about the treatments they are receiving. Also, a different study discovered that relationship marketing might increase a client's confidence in the reliability of the services offered (Wongsansukcharoen, 2022; Salem, 2021).

Relationship marketing, in the words of Afifi and Amini (2021), is marketing done with the intention of establishing long-term influence or trust with regard to clients, distributors, suppliers, or other partners in the market structure. The findings of this study have a big impact on academics and corporate practices. It is clear from a business standpoint that various degrees of relationship marketing can lead to various degrees of customer trust. This conclusion implies that managers should concentrate on relationship marketing practices to increase and improve customer trust and, thereby, boost brand image. In order for the hospital to accomplish this, they must deliver on the services they have committed to supply to customers. For instance, hospitals should concentrate on enhancing the quality of their services and train their employees on how to treat consumers with compassion, depending on the circumstances. As a result, hospitals can enhance connections through feedback to eliminate doctor-patient separation and enable two-way communication.

### *H3: Patient Trust Has a Positive Significant Effect on Hospital Brand Image*

The standardized coefficient of the impact of patient trust on hospital brand image (HBIM <--- TR) was 0.250***, which explains that patient trust (TR) has a significant positive impact on trust (HBIM). This outcome is consistent with the earlier research, which discovered that developing provider-customer-centeredness and building trust through good communication had a favorable and significant impact on brand image (Lian & Yoong, 2017). In today's extremely competitive global market, maintaining a trusted and long-lasting relationship with clients is essential for a brand to succeed. Consumers' trust in service providers refers to their anticipation that providers will fulfill their promises in a trustworthy and reliable manner (Al-Adwan et al., 2020). The brand image of any business or organization is greatly influenced by trust. Customers are more willing to interact with and spread the word about a brand they trust. Yet, a lack of trust can lead to unfavorable perceptions and a tarnished brand image. Building deep ties with customers is the goal of relationship marketing, which can help generate trust and develop a favorable brand image. When customers trust a brand, they are more inclined to recommend it to others, leading to word-of-mouth marketing and increased brand recognition. In addition, devoted customers are more likely to stick with a company even when presented with alternatives. Brands can set themselves apart from rivals by prioritizing relationship marketing and trust-building.

### *H4: Patient Trust Mediates the Link between Relationship Marketing and Hospital Brand Image*

The mediating result of relationship marketing on hospital brand image (HBIM ← PTR← RMRKT) was 0.055***; this explains that trust (TR) mediates the link between relationship marketing (RMRKT) and hospital brand image (HBIM). Building great customer relationships, providing outstanding customer service, and engaging in relationship marketing techniques can all contribute to a positive brand image by fostering a sense of trust among customers and the company. Prior research has revealed that relationship marketing techniques have a favorable influence on brand trust, leading to a possible favorable impact on brand devotion, which can lead to the company having a positive brand image (Ebrahim, 2020; Mohd Isa, Lim, & Chin, 2019; Shahin Sharifi & Rahim Esfidani, 2014). Consequently, enhancing a brand's image and reputation using relationship marketing techniques can be successful. Businesses can develop a positive brand image and reputation that can result in more devoted customers and higher sales by putting a priority on relationship marketing techniques that foster trust.

## Theoretical Implication

This study provides numerous theoretical contributions. Within hospitals, there hasn't been much research on the connection between relationship marketing and hospital brand image through the lens of trust. First, this study focuses on Nigeria's healthcare sector while integrating relationship marketing, trust, and hospital brand image. Our research advances knowledge of the role that trust plays in the healthcare sector's interaction between relationship marketing and hospital brand perception. We demonstrate how trust increases the beneficial impact of relationship marketing on hospital brand image. Our findings support the hypothesis that relationship marketing affects hospital brand image in terms of theoretical contributions. We suggest a novel way to increase hospital brand image using relationship marketing, while trust serves as a mediator.

## Managerial Implication

The findings of this study suggest some managerial implications that need to be considered. In order to determine hospital brand image through trust, relationship marketing is first assessed using an interactive and complete model provided by the research. The study's conclusions indicate that for public hospitals to continue operating successfully, a focus on patients' impressions of relationship marketing and trust is essential. Hospital management and marketing teams should place a high priority on forging deep emotional bonds with patients to promote a favorable hospital image. In order to enhance patient experiences, hospitals can also send staff members to relationship marketing training sessions. Hospital staff members and professionals should uphold a high standard of honesty and promote a great reputation. These initiatives may help improve the hospital's reputation and establish enduring relationships with patients.

## Limitations and Further Research

The study's only reliance on quantitative data forms its limitation. The particular sector, area, and sampling technique and sample size are also connected with further constraints. Therefore, the results cannot be broadly applicable to many businesses and nations. For future studies, we suggest using the mix method to examine the impact of relationship marketing on hospital brand image in order to address this issue. Additionally, it would be advantageous to broaden the analysis's scope to include diverse industries and nations using a representative sample. The model should also take into account additional factors, including patient value, commitment, and loyalty.

## Editors Note

The investigation into the mediating effect of patient trust on the relationship between relationship marketing and hospital brand image in public hospitals in Nigeria is of paramount importance for sustainability within the healthcare sector. Relationship marketing strategies play a pivotal role in fostering positive interactions and building long-term relationships between healthcare providers and patients. By examining how these strategies influence patient trust and, subsequently, hospital brand image, this research offers valuable insights into enhancing patient perceptions and loyalty towards public healthcare institutions. A strong hospital brand image not only attracts patients but also instills confidence and credibility in the healthcare services provided, ultimately contributing to the sustainability of public hospitals in Nigeria. Furthermore, understanding the mediating role of patient trust sheds light on the underlying mechanisms through which relationship marketing practices impact hospital brand image, enabling healthcare administrators to develop targeted interventions that foster trust, enhance patient experiences, and promote the long-term viability of public healthcare systems. Therefore, this research has the potential to inform strategic decision-making processes aimed at improving service quality, patient satisfaction, and overall sustainability within the Nigerian healthcare landscape.

## References

Abid, M. F., Siddique, J., Gulzar, A., Shamim, A., Dar, I. B., & Zafar, A. (2023). Integrating the commitment-trust theory to gauge customers loyalty in riding services. *Journal of Promotion Management, 29*(3), 305–337.

Afifi, I., & Amini, A. (2021). Factors affect to relationship marketing for creating customer loyalty in hospital services business. *ASEAN Marketing Journal, 10*(2), 2.

Al-Adwan, A. S., Kokash, H., Adwan, A. A., Alhorani, A., & Yaseen, H. (2020). Building customer loyalty in online shopping: The role of online trust, online satisfaction and electronic word of mouth. *International Journal of Electronic Marketing and Retailing*, *11*(3), 278–306.

Allahham, L., Mouselli, S., & Jakovljevic, M. (2022). The quality of Syrian healthcare services during COVID-19: A HEALTHQUAL approach. *Frontiers in Public Health*, *10*, 970922.

Ashton, A. S., Scott, N., Solnet, D., & Breakey, N. (2010). Hotel restaurant dining: The relationship between perceived value and intention to purchase. *Tourism and Hospitality Research*, *10*(3), 206–218.

Atulkar, S. (2020). Brand trust and brand loyalty in mall shoppers. *Marketing Intelligence & Planning*, *38*(5), 559–572.

Bernarto, I., Berlianto, M. P., Meilani, Y. F. C. P., Masman, R. R., & Suryawan, I. N. (2020). The influence of brand awareness, brand image, and brand trust on brand loyalty. *Jurnal Manajemen*, *24*(3), 412–426.

Berne-Manero, C., & Marzo-Navarro, M. (2020). Exploring How Influencer and Relationship Marketing Serve Corporate Sustainability. *Sustainability*, *12*(11), 4392. <https://doi.org/10.3390/su12114392>

Biddle, L., Ziegler, S., Baron, J., Flory, L., & Bozorgmehr, K. (2022). The patient journey of newly arrived asylum seekers and responsiveness of care: A qualitative study in Germany. *Plos One*, *17*(6), e0270419.

Boafo, N. D., Agyapong, F., Asare, P., & Amponsah, G. (2020). *The balance between corporate identity and corporate image and its impact on marketing of universities in Ghana.*

Campbell, M. C., Inman, J. J., Kirmani, A., & Price, L. L. (2020). In times of trouble: A framework for understanding consumers' responses to threats. *Journal of Consumer Research*, *47*(3), 311–326.

Cham, T. H., Lim, Y. M., Aik, N. C., & Tay, A. G. M. (2016). Antecedents of hospital brand image and the relationships with medical tourists' behavioral intention. *International Journal of Pharmaceutical and Healthcare Marketing*, *10*(4), 412–431.

Chen Y., Yan X., Fan W., & Gordon, M. (2015). The joint moderating role of trust propensity and gender on consumers' online shopping behavior, *Computers in Human Behavior*, *43*, 272–283. <http://dx.doi.org/10.1016/j.chb.2014.10.020>

Chen, W. J., & Chen, M. L. (2014). Factors affecting the hotel's service quality: Relationship marketing and corporate image. *Journal of Hospitality Marketing & Management*, *23*(1), 77–96.

Colgan, D. D., Eddy, A., Aulet-Leon, M., Green, K., Peters, B., Shangraw, R., … Oken, B. (2021). Compassion, communication, and the perception of control: A mixed methods study to investigate patients' perspectives on clinical practices for alleviating distress and promoting empowerment during awake craniotomies. *British Journal of Neurosurgery*, 1–12.

Darmawan, D., & Grenier, E. (2021). Competitive advantage and service marketing mix. *Journal of Social Science Studies (JOS3)*, *1*(2), *38*, 75–80.

Du, L., Xu, J., Chen, X., Zhu, X., Zhang, Y., Wu, R., … Zhou, L. (2020). Rebuild doctor—patient trust in medical service delivery in China. *Scientific Reports*, *10*(1), 21956.

Ebrahim, R. S. (2020). The role of trust in understanding the impact of social media marketing on brand equity and brand loyalty. *Journal of Relationship Marketing*, *19*(4), 287–308.

Ekman, N., Moons, P., Taft, C., Boström, E., & Fors, A. (2022). Observable indicators of person-centred care: An interview study with patients, relatives and professionals. *BMJ Open*, *12*(4), e059308.

Elizar, C., Indrawati, R., & Syah, T. Y. R. (2020). Service quality, customer satisfaction, customer trust, and customer loyalty in service of paediatric polyclinic over private h hospital of East Jakarta, Indonesia. *Journal of Multidisciplinary Academic*, *4*(2), 105–111.

Fazal-e-Hasan, S., Mortimer, G., Lings, I., & Kaur, G. (2020). How gratitude improves relationship marketing outcomes for young consumers. *Journal of Consumer Marketing*, *37*(7), 713–727.

Febriyantoro, M. T. (2020). Exploring YouTube marketing communication: Brand awareness, brand image and purchase intention in the millennial generation. *Cogent Business & Management*, *7*(1), 1787733.

Gambarov, V., Sarno, D., Hysa, X., Calabrese, M., & Bilotta, A. (2017), The role of loyalty programs in healthcare service ecosystems. *The TQM Journal*, *29*(6), 899–919. <https://doi.org/10.1108/TQM-02-2017-0019>

Geese, F., Willener, R., & Spichiger, E. (2021). Patients' and partners' experiences with prostate cancer and advanced practice nurse counselling. *International Journal of Urological Nursing*, *15*(2), 82–90.

Hair, J. F., Anderson, R. E., Tatham, R. L., & Black, W. C. (1998). *Multivariate data analysis* (5th edn). Upper Saddle River, NJ: Prentice Hall International.

Han, H., & Hyun, S. (2015). Customer retention in the medical tourism industry: Impact of quality, satisfaction, trust, and price reasonableness, Tourism Management, *46*, 20–29. <https://doi.org/10.1016/j.tourman.2014.06.003>

Hayati, S., Suroso, A., Suliyanto, S., & Kaukab, M. (2020). Customer satisfaction as a mediation between micro banking image, customer relationship and customer loyalty. *Management Science Letters*, *10*(11), 2561–2570.

Heffernan, T., O'Neill, G., Travaglione, T., & Droulers, M. (2008). Relationship marketing: The impact of emotional intelligence and trust on bank performance. *International Journal of Bank Marketing*, *26*(3), 183–199.

Irizarry, T., DeVito Dabbs, A., & Curran, C. R. (2015). Patient portals and patient engagement: A state of the science review. *Journal of Medical Internet Research, 17*(6), e148.

Kholis, N., & Ratnawati, A. (2021). The effects of customer engagement and brand trust on brand loyalty: A case study of BPJS healthcare participants in Indonesia. *The Journal of Asian Finance, Economics and Business, 8*(11), 317–324.

Kim, S. T., Lee, H. H., & Hwang, T. (2020). Logistics integration in the supply chain: A resource dependence theory perspective. *International Journal of Quality Innovation, 6*, 1–14.

Konuk, F. A. (2019). Consumers' willingness to buy and willingness to pay for fair trade food: The influence of consciousness for fair consumption, environmental concern, trust and innovativeness. *Food Research International, 120*, 141–147.

Konuk, F. A. (2021). The moderating impact of taste award on the interplay between perceived taste, perceived quality and brand trust. *Journal of Retailing and Consumer Services, 63*, 102698.

Lacap, J. P. G., Cham, T. H., & Lim, X. J. (2021). The influence of corporate social responsibility on brand loyalty and the mediating effects of brand satisfaction and perceived quality. *International Journal of Economics & Management, 15*(1), 69–87.

Lee, C. (2021). Patient loyalty to health services: The role of communication skills and cognitive trust. *International Journal of Healthcare Management, 14*(4), 1254–1264.

Lian, S. B., & Yoong, L. C. (2017). The effectiveness of strategic relationship marketing: Exploring relationship quality towards customer loyalty. *International Business Research, 10*(12), 159–166.

Liao, H., Wu, X., Mi, X., & Herrera, F. (2020). An integrated method for cognitive complex multiple experts multiple criteria decision making based on ELECTRE III with weighted Borda rule. *Omega, 93*, 102052.

Limbu, Y. B., Pham, L., & Mann, M. (2020). Corporate social responsibility and hospital brand advocacy: Mediating role of trust and patient-hospital identification and moderating role of hospital type. *International Journal of Pharmaceutical and Healthcare Marketing, 14*(1), 159–174.

Malhotra, P. (2019). *Building trust to create crowdsourcing acceptance*. London: British Academy of Management.

Mashayekhi, F., Rajabpour, E., Ahmadi, H., & Esmaeilpour, M. (2022). The effect of job crafting on patients satisfaction: With the role of moderate of quality of health services (Case study: Persian Gulf Hospital in Bushehr). *Journal of Advertising and Sales Management, 3*(4), 40–54.

Ngoma, M., & Ntale, P. D. (2019). Word of mouth communication: A mediator of relationship marketing and customer loyalty. *Cogent Business & Management, 6*(1), 1–20.

Nguyen, T. M., Quach, S., & Thaichon, P. (2022). The effect of AI quality on customer experience and brand relationship. *Journal of Consumer Behaviour, 21*(3), 481–493.

Özkan, P., Süer, S., Keser, İ. K., & Kocakoç, İ. D. (2020). The effect of service quality and customer satisfaction on customer loyalty: The mediation of perceived value of services, corporate image, and corporate reputation. *International Journal of Bank Marketing, 38*(2), 384–405.

Ponathil, A., Khasawneh, A., Byrne, K., & Chalil Madathil, K. (2020). Factors affecting the choice of a dental care provider by older adults based on online consumer reviews. *IISE Transactions on Healthcare Systems Engineering, 11*(1), 51–69.

Ryan, M. (2020). In AI we trust: Ethics, artificial intelligence, and reliability. *Science and Engineering Ethics, 26*(5), 2749–2767.

Salem, S. F. (2021). Do relationship marketing constructs enhance consumer retention? An empirical study within the hotel industry. *SAGE Open, 11*(2), 21582440211009224.

Setiawan, E., Wati, S., Wardana, A., & Ikhsan, R. (2020). Building trust through customer satisfaction in the airline industry in Indonesia: Service quality and price fairness contribution. *Management Science Letters, 10*(5), 1095–1102.

Shahin Sharifi, S., & Rahim Esfidani, M. (2014). The impacts of relationship marketing on cognitive dissonance, satisfaction, and loyalty: The mediating role of trust and cognitive dissonance. *International Journal of Retail & Distribution Management, 42*(6), 553–575.

Solangi, B., Talpur, U., Shaikh, S. S., Mushatque, T., & Channa, M. A. (2019). Relationship marketing as an orientation to customer retention: Evidence from banks of Pakistan. *International Business Research, 12*(7), 133–140.

Sousa, B. M., & Alves, G. M. (2019). The role of relationship marketing in behavioural intentions of medical tourism services and guest experiences. *Journal of Hospitality and Tourism Insights, 2*, 224–240.

Stumpp, N. E., & Sauer-Zavala, S. (2022). Evidence-based strategies for treatment personalization: A review. *Cognitive and Behavioral Practice, 29*(4), 902–913.

Suryadi, S., Pasinringi, S. A., & Kadir, A. R. (2022). The effect of customer relationship marketing and customer value on customer loyalty at Syekh Yusuf Hospital, Gowa Regency. *Journal of Asian Multicultural Research for Economy and Management Study, 3*(2), 21–28.

Tran, V. D., & Vu, Q. H. (2019). Inspecting the relationship among E-service quality, E-trust, E-customer satisfaction and behavioral intentions of online shopping customers. *Global Business & Finance Review (GBFR), 24*(3), 29–42.

Veselova, A. (2022). The Role of Relationship Marketing in the Context of Building Customer Satisfaction. *Scientific Programme Committee*, 262–269.

Warren, C., Carter, E. P., & McGraw, A. P. (2019). Being funny is not enough: The influence of perceived humor and negative emotional reactions on brand attitudes. *International Journal of Advertising*, *38*(7), 1025–1045.

Wolf, E. J., Harrington, K. M., Clark, S. L., & Miller, M. W. (2013). Sample size requirements for structural equation models: An evaluation of power, bias, and solution propriety. *Educational and Psychological Measurement*, *73*(6), 913–934.

Wolter, J. S., Bock, D. E., Hopkins, C. D., & Giebelhausen, M. (2022). Not the relationship type? Loyalty propensity as a reason to maintain marketing relationships. *Journal of the Academy of Marketing Science*, *50*(5), 1052–1070.

Wongsansukcharoen, J. (2022). Effect of community relationship management, relationship marketing orientation, customer engagement, and brand trust on brand loyalty: The case of a commercial bank in Thailand. *Journal of Retailing and Consumer Services*, *64*, 102826.

Zewdu, F. (2022). *The effect of customer relationship marketing on customer loyalty: The mediating role of customer satisfaction in selected banking sector in the case of Bahir Dar City*, Doctoral dissertation.

ALEEM ABDUL-KAREEM AND STEPHEN TAIWO ONIFADE

# 5 The Effect of Organizational Culture on Individual Creativity: An Empirical Investigation of Packaging Firms in Büsan Industrial Site of Central Anatolia Province, Konya-Türkiye

ABSTRACT
A crucial aspect of organizational setup that creates distinctions among members of different organizations can be attributed to organizational culture. The kind of culture that exists in a business setting has the propensity of affecting the attitudes, perceptions, actions and thoughts of individuals towards creativity. In this study, we therefore assess the effect of organizational culture on creativity of employees of packaging firms operating in Büsan organized industrial site in Konya. Structured questionnaires were administered to a total of one hundred and twenty-four (124) participating employees to source primary data for the empirical analysis in the study. The analysis encompasses a comprehensive application of regression and correlation techniques on the field data. Our findings divulge significant positive effect of organizational culture on individual's creativity levels. Furthermore, the different dimensional views of organizational culture (Klan, Adhocracy, Market and Hierarchy Culture) reflect significant positive effect on individual's creativity. Based on these findings, adequate resources and special attention to organizational culture are essentially advocated for fostering individual creativity towards attaining organizational missions and visions.

Keywords: Culture, creativity, organizational culture, individual creativity, Türkiye

JEL Codes: M14, O34

## Introduction

Organizational culture remains a part of the issues in academic research, education and in management practice. The reason is that the cultural dimension is central in all aspects of organizational life (Alvesson, 2012). The intangible aspects of an organization such as its culture are as important as the tangible aspects. The factors surrounding organizational culture are not only critical to understanding organizations but can also help decipher their success rates or otherwise (Demir, Unnu, & Erturk, 2011). This calls for the need for management to pay keen attention to the kind of culture that exists in an organization. Management should focus on creating a strong and preferred culture in an organization. Strong culture essentially shapes the attitudes of employees to be more productive, committed and creative. The culture of an organization defines the set of norms, values, traditions, symbols and beliefs that are held by members of that organization. Organizational culture specifies how an organization will interact with its key stakeholders such as customers, suppliers, employees, creditors, competitors among others. According to Alvesson (2012) in most contemporary organizations, culture receives a lot of attention and is seen as a crucial component. The actions and reactions, orientations as well as feelings of workers are often influenced by cultural nature of an organization regardless of whether cultural issues are sufficiently addressed or taken less seriously by the management.

The business world requires a more understanding of how organizational culture affects individual creativity. The real foundation for innovative ideas is creativity, thus, the latter is the true bedrock that puts any organization in an advantageous position as far as competition is concerned in business. This is why it has been noted that different firms are now keen on advancing their competitiveness via the creativity of their individual employees (Muñoz-Doyague & Nieto, 2012). Besides, it has further been noted that more and more firms are seeking after creative ideas to boost customers satisfaction as well as brand loyalty (Gül, 2019).

Hence, we attempt to empirically investigate the impact of organizational culture on individual employee creativity of packaging firms operating in Büsan Industrial Site of Central Anatolia Province, Konya-Türkiye. Our study helps to bridge some gaps in the relevant areas within literature. Currently, there is a handful of empirical studies specifically assessing the nexus of individual employee creativity with organizational culture. The study extends its focus to review literature on the understudied factors. The subsequent sections after the literature review contain the methodology part, while the analysis and the ensuring findings were thereafter explained. Later on, the overall conclusions and managerial implications were drawn, and other research limitations were put forward.

## Review of Literature

### *Contextualizing Organizational Culture*

To present a vivid definition of organizational culture, there is a need to firstly reflect on the meaning of the word culture. Culture is a very broad concept with a wide range of definitions. Culture encompasses set of unwritten rules of a social environment. It is general conditioning of the mind that create distinctions among people(s), members of a community, or among group(s) (Hofstede, Hofstede, & Minkov, 2010). Cultural values are learned and cannot be transmitted through hereditary. Culture is symbolic in nature, shared by a group of people and passes from one generation to the other. It is integrated, dynamic and has a structure (Luthans & Doh, 2017). Right from the early 1980s, the ideas of organizational culture has being gaining appreciation (Hofstede et al., 2010). Organizational culture can be traced to Organizational theory and has many different and competing definitions. It is seen as a complex set of values, beliefs, assumptions and can also include symbols that explain the why firms or businesses operate the way they do (Damaraju, Barney, & Dess, 2021). It can also be viewed as the emotions, values, beliefs, assumptions, norms, perceptions, symbols and attitudes shared by the members of a group that are different from other groups. It is mostly reflected and enshrined in the principles that guide members of a workplace and passed unto new employees via certain approaches including symbolism, the use of myths and stories relating to expected behavior in the organization (Schneider, Ehrhart, & Macey, 2013). It essentially reflects an arrangements of shared ideologies that distinguish the members of an organization that hold them from the others (Robbins & Judge, 2016).

When the organizational culture is evaluated as a whole; It consists of functional elements such as cultural values, basic assumptions, norms, attitudes, traditions, symbols, beliefs, language, myths, stories, rituals, taboo and Organizational heroes (Schein & Schein, 2016). The essence of an organizational culture can be seen in seven (7) main features to include, innovation and risk taking, attention to detail, orientation towards outcome, orientation towards people, collaboration and team orientation, aggressiveness and finally stability (Borg, Groenen, Jehn, Bilsky, & Schwartz, 2011; Robbins & Judge, 2016).

This section provides information about the classification of organizational culture. It is classified in different ways by researchers on this topic. The classification form into strong and weak cultures is mostly referred to in the literature (Schneider et al., 2013). Members of a strong culture usually hold their shared values very tightly. On the contrary, in a weak culture, the organization's key values and set of norms are not broadly shared and firmly held. In general, a strong organizational culture is accepted as the preferred culture (Grover, Tseng, & Pu, 2022). Harrison (1995) provided four (4) major classifications ranging from power culture to role culture, task culture, and individual culture. Handy (1996) supported Harrison's classification and added bureaucratic and democratic cultures. Pheysey (2002) also included role and power cultures in her classification but added achievement and support cultures too. Cameron and Quinn (2011), proposed four types of organizational cultures: clan, market, adhocracy, and hierarchy cultures. This classification emanated from their competitive values model. This model then forms the basis of the Organizational Culture Assessment Scale. The classification of Cameron and Quin is explained below.

**Clan Culture:** It denotes a friendly and comfortable working environment in which the workforce is perceived as an extended family (Fan, Li, Xue, & Yu, 2023). This culture type distinguishes itself from others has it primarily focuses on establishing and strengthening team spirit and propel stronger cohesion among members toward boosting their organizational allegiance. "Clan" is specifically used for this type of culture because of similarity to an organization in a family settings (Cameron & Quinn, 2011).

**Market Culture:** Organizations that adopt this kind of culture focus on the external environment and try to have an important place in the market. The leaders in these organizations are determined and eager to achieve long-term goals. Employees are motivated and rewarded according to their contributions to the organization (Low, Abdul-Rahman, & Zakaria, 2020). Organizations characterized by a market culture emphasize transactions and relationships with external parties such as customers, suppliers, creditors, regulators among others. The purpose of these organizations is to increase their market share, maximize profits, increase market power and secure more customers (Cameron & Quinn, 2011).

**Adhocracy Culture**: this is more like a temporary institution, which is disbanded whenever its tasks are completed and reformed as soon as newer ones come up. This type of organizational culture gives a lot of opportunities to individuals to be creative, adaptive and flexible in working towards the Organizational goals (Aktaş et al, 2011). Organizations having such culture are characterized with dynamism, adaptability and creativity. Employees are encouraged to think and act freely. They are also expected to develop and utilize their entrepreneurial skills towards achieving the Organizational objectives. The term itself is from two words; "ad" and "hoc" which implies something temporary, specialized and dynamic (Cameron & Quinn, 2011).

**Hierarchy Culture:** It describes a culture characterized by formal rules, regulations and policies in which interpersonal relationships are well defined and coordinated. It is a culture where orders, procedures, status and accountability are at the forefront, and close monitoring and evaluation is exercised (Valmohammadi

& Roshanzamir, 2015). Maintaining a smoothly running organization is important in this type of organizational culture. Many organizations are concerned about their stability and efficiency especially in the long run, so, official guidelines are set to help keep the organization together (Cameron & Quinn, 2011).

## Individual Creativity

One of the significant resources used in solving the problems of organizations and adjust to unstable business environment is individual employee (Oldham, 2003). This is because of the knowledge, talent and skills individual employees possess. In this section, a brief description of the meaning of creativity is provided before delving deeper into the intricacies of individual creativity. Creativity is basically the creation of novel, suitable and useful ideas in any domain of human activity. The domain could range from one field to the other. It could be from science, to arts, to business, to education or to everyday life. The ideas must be novel. Thus, it must be similar to what was done before, but the ideas cannot be merely bizarre. The ideas must be suitable and relevant in solving the current problem or harnessing the opportunity identified (Walia, 2019). In innovation, creativity is regarded as its first step. Creativity is described as the manager's most powerful and effective weapon due to its propensity to effectively solve business and customers problems by providing effective solutions to these problems. Therefore, it would be correct and reasonable to consider creativity as the key and foundation to Organizational success (Glăveanu & Beghetto, 2021). Based on traditional approach, creativity is something done by creative people. In general terms, creativity can be examined in terms of individual creativity and Organizational creativity. The focus of the present study will be on individual/employee creativity.

Individual creativity refers to the activities and actions taken by employees individually in an Organizational setting in order to improve their capabilities of inventing something considered novel, meaningful and appropriate in solving Organizational problem (Bharadwaj & Menon, 2000). In order for an organization to be innovative and attain sustainable competitive advantage, its employees need to consistently engage in creativity. Thus, they have to be actively involved in their jobs and produce new and suitable processes, products and approaches. Creative employees serve as role models to others in the organization and their new ideas are transferable to others for their own usage and development (Shalley & Gilson, 2004). As individual creativity is an important source of Organizational innovation and competitive advantage, organizations are increasingly seeking to foster individual creativity (Zhou & George, 2001). Individual creativity describes how complex problems in business operations and activities are resolved by generating workable ideas (Arif, Qaisar, & Kanwal, 2022). Depending on how creative individuals are, they can propose different solutions to the same problem. Individual creativity involves discovering relations that exist between things or objects, looking at those objects from different new perspectives and creating new combinations from the previous concepts (Taggar, 2002).

Creative individuals have the ability to discover new ways of using existing methods or equipment or produce novel but practicable work-related ideas. Creative individuals do not just developed the ideas to tackle problems, they also strategize on how to ensure that those ideas are successfully implemented to reach the desired goals (Gumusluoglu & Ilsev, 2009). Creative individuals are more likely to identify consumers' covert needs and effectively solve their problems in order to result in superior firm performance. Studies on individual creativity show that creative individuals possess traits that can be personal or common to a group. For instance, these people largely have self-confidence, not afraid of complexity, tolerance of ambiguity and intuitive (Çekmecelioğlu & Günsel, 2013). According to (Amabile, 2017) individual creativity basically or essentially entails three components. These are expertise, creative-thinking skills and intrinsic task motivation. These components are essential for a meaningful individual creativity to take place. There are other factors that have the tendency of either positively or negatively affecting the development of individual creativity. Perry-Smith (2006) emphasizes on social ties or relationships as significant factor affecting individual creativity. Individuals with weak social ties are more creative than those with strong social ties and the working atmosphere can also play some roles. When there exist good cooperation, support and mutual trust among members of a team, members tend to be more confident and try to develop new approaches to do their work (Meng, Cheng, & Guo, 2016). The kind of relationship leaders have with employees can directly or indirectly affect individual creativity. Employees may be motivated or demotivated from willing to take risks that will culminate into novel ideas (Zhou & George, 2003). The level and quality of the relationships that employees establish with their superiors and colleagues have an important impact on their creativity (Muñoz-Doyague & Nieto, 2012).

## Review of the Empirical Literature

For those organizations that can sustain a strong and "preferred" culture, they are more likely to reap the benefits of better performance, higher creativity, and they can be better positioned for competitive advantage (Demir et al., 2011). Therefore, it has been noted that it is more challenging to create culture that can help convert employees creativity to desired organizational gains from the perspective a higher market share and better customer satisfaction (Csikszentmihalyi & Sawyer, 2014). The development of creative ideas in organizations is linked to organizational culture. Organizations that create an organizational culture that fosters creativity will unearth the creativity of their employees and will become businesses that grow rapidly in today's competitive environment (Telli, 2016). Organizational-level factors such as culture, rewards have both direct and indirect influences on individual creative behaviors (Agars, Kaufman, Deane, & Smith, 2012). It is important for employees' beliefs to align with the organizational goals to have a creative culture. For instance, some workers belief that new and challenging times offers some opportunity for creativity, as such, they have the readiness to take risks in anticipation of future success (Tesluk, Farr, & Klein, 1997).

Telli (2016) revealed that organizational culture has a positive effect on individual creativity in a study conducted on 170 nurses in Kahramanmaraş Necip Fazıl City Hospital. In a study conducted by Ogbeibu, Senadjki, and Gaskin (2018), it was found that adhocracy organizational culture has a significant positive effect on employee creativity. The results of the study also revealed that market and clan cultures have significant negative effects on employee creativity, and hierarchy culture does not really affect creativity levels. A significant positive effect of organizational culture on employees' creative behavior was shown in a research conducted by Gupta (2011) on 128 executives from six different organizations.

## Research Methodology

### *Research Purpose and Sample*

The purpose of this study is to examine the effect of organizational culture on creativity of employees of packaging firms operating in Büsan organized industrial site of Central Anatolia Province, Konya-Türkiye. This research was carried out on packaging firms operating in Büsan Industrial Site of Konya, Türkiye. As at the time of conducting this research, there were a total of eleven (11) packaging firms in the aforementioned industrial site. Nine (9) out of eleven (11) firms agreed to participate in the study. The firms that took part in the study are; Ambalaj Dünyası, Sen Kardeşler Ambalaj, Sedefsan Ambalaj, Efofleks Plastik ambalaj, YSN Pak ambalaj, En Cup Ambalaj, Ekvator Kağıt Ambalaj, MFS ambalaj ve Özkan Ambalaj. The population of this research is made up of employees of these packaging firms. Questionnaires were administered as the main instrument to gather the required data for this study. A total of one hundred and twenty-four (124) employees participated in the research. Simple random sampling technique was adopted in selecting the sample size.

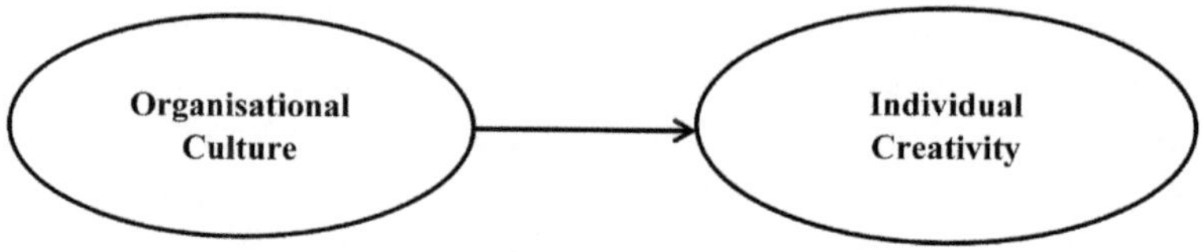


Figure 5.1: Conceptual framework

## Research Model and Hypotheses

We modeled the links among the understudied factors schematically as seen in Figure 5.1 below.

Based on the purpose of the study, review of the literature and conceptual framework presented above, the following hypotheses were proposed.

H1: Organizational culture positively affects individual creativity.

H1a: Clan culture positively impacts individual creativity.

H1b: Adhocracy culture positively impacts individual creativity.

H1c: Market culture positively impacts individual creativity.

H1d: Hierarchy culture positively impacts individual creativity.

## Measurement Tools

A questionnaire approach was employed to gather the necessary data for this study. The research questionnaires consisted of three main parts with a total of 43 questions. The first part was meant to gather data on the demographic characteristics of the respondents. The second part was devoted to the independent variable (organizational culture). The third part contained questions about the dependent variable (individual creativity). Previous researchers who conducted similar and related studies developed the scales used in this study. Therefore, all the scales used have already been subjected to validity and reliability tests and have proven to be valid and reliable for empirical studies. Information about the scales adapted for this study is given below.

A scale developed by Cameron and Quinn (1999) and translated into Turkish by İşcan and Timuroğlu (2007) was used to measure organizational culture. The scale has a structure consisting of 24 items and 4 dimensions. Out of these 24 items, the first six items measure clan culture, the second six items evaluate adhocracy, the third six assess market and the last six items measure hierarchy culture. All 24 items were adapted into this study and measured using the 5-point Likert scale starting from 1 = Strongly Disagree to 5 = Strongly Agree.

The individual creativity scale developed by Zhou and George (2003) and adapted into Türkiye's context by Gumusluoglu and Ilsev (2009) in their studies was used to evaluate the creativity of employees. It is a unidimensional scale and consists of 13 items. All 13 items were adapted into this study and measured using the 5-point Likert type starting from 1 = Strongly Disagree to 5 = Strongly Agree.

## Data Analysis and Findings

Descriptive research design combined with both quantitative and qualitative analyzes were employed in this study. The data gathered from the participants were analyzed using SPSS software package. The demographic information of the data collected as seen in Table 5.1 consist of gender information, age and marital information, and the necessary information on education, title of job title, and the level of respondents' experience in the jobs. The results are presented on Table 5.1.

Looking at the survey outcomes on Table 5.1, we can deduce that a sizeable proportion of the respondents are male. The males account for about 71 % while a little lower percentage to this (60.5 %) fall in the category of the married. As for the age distribution, the largest group of the age distribution is the age bracket (21–30), followed by those in the (31–40) groups with 57.3 % and 24.2 % respectively. As for educational attainments,

Table 5.1: Demographic data

| **Variables** | | **Frequency** | **Percent (%)** |
|---|---|---|---|
| Gender | Male | 88 | 71.0 |
| | Female | 36 | 29.0 |
| | Total | 124 | 100 |
| Marital Status | Married | 75 | 60.5 |
| | Single | 45 | 39.5 |
| | Total | 124 | 10.0 |
| Age | Below 21 years | 11 | 8.9 |
| | 21–30 | 71 | 57.3 |
| | 31–40 | 30 | 24.2 |
| | 41–50 | 8 | 6.5 |
| | 51–60 | 2 | 1.6 |
| | Above 60 years | 2 | 1.6 |
| | Total | 124 | 100 |
| Educational Level | Primary Sch. | 1 | 0.8 |
| | Middle Sch. | 6 | 4.8 |
| | Senior High Sch. | 24 | 19.4 |
| | Vocational High Sch. | 7 | 5.6 |
| | Associate Degree | 17 | 13.7 |
| | Degree | 55 | 44.4 |
| | Master's Degree | 12 | 9.7 |
| | Doctorate | 2 | 1.6 |
| | Total | 124 | 100 |
| Job Position | Owner/Partner | 1 | 0.8 |
| | Gen. Manager/Assistant Gen. Manager. | 4 | 3.2 |
| | Supervisor/Foreman | 18 | 14.5 |
| | Department/Unit Manager/ Assistant Department Manager | 15 | 12.1 |
| | Employee/Officer | 76 | 61.3 |

Table 5.1: Continued

| **Variables** | | **Frequency** | **Percent (%)** |
|---|---|---|---|
| | Other | 10 | 8.1 |
| | Total | 124 | 100 |
| Overall Years of Work Experience | Below 1 year | 10 | 8.1 |
| | 1–5 | 50 | 40.3 |
| | 6–10 | 28 | 22.6 |
| | 11–15 | 17 | 13.7 |
| | 16–20 | 5 | 4.0 |
| | Above 60 years | 14 | 11.3 |
| | Total | 124 | 100 |

workers with bachelors' degree are the topmost group, while the high school certificate holders are the next, thus, implying about 44.4 % and 19.4 % respectively for each group. When we look at the job titles, around 61.3 % claimed to be ordinary employees while just around 14.5 % work as supervisors. Lastly, around 40.3 % of the workers have the work experience of between at least a year or about five years.

### *Reliability and Validity Tests*

Cronbach Alpha value was used to determine the reliability levels of the scales. Table 5.2 displays the reliability coefficients of organizational culture and individual creativity scales.

When examining Table 5.2, we can confidently say that adequate scales have been adopted since the Cronbach alpha co-efficient values exceed the 0.7 conventional level thus, confirming the scales' consistency as well as internal stability.

In order to determine the validity of the scales used in this study, varimax rotation principal component (exploratory) factor analysis was conducted on the scales. The results of factor analysis on organizational culture corroborate the four-factor structure of the scale. The factor analysis

Table 5.2: The test of reliability

| Scale | The Measurement Interval | The Number of Items | The Cronbach Alpha Value |
|---|---|---|---|
| Organizational Culture | 5-point Scale | 24 | 0.961 |
| Individual Creativity | 5-point Scale | 13 | 0.933 |
| **Organizational Culture Dimensions**<br>For Clan Culture<br>For Adhocracy Culture<br>For Market Culture<br>For Hierarchy Culture | 5-point Scale | 6<br>6<br>6<br>6 | 0.879<br>0.909<br>0.887<br>0.866 |

resulted in four dimensions, clan, adhocracy, market and hierarchy culture. The eigenvalues of the four dimensions (the clan, adhocracy, market and hierarchy culture) are 12.861, 1.499, 1.057 and 1.035 respectively with a total cumulative variance of 64.239 %, which is more than the recommended 50 %. The factor loadings of each item on the scale are greater than 0.50 and their eigenvalues are beyond the ideal value of 1 (Shrestha, 2021). The KMO value is 0.930 and the Bartlett test ($X^2$) result of 2,220.390 with $p < 0.001$.

Outcomes of the factor analysis further confirm the one-dimensional structure of the individual creativity scale. The eigenvalue of this single dimension is 7.266 with a total cumulative variance of 51.901 %. The KMO value is 0.886 and the Bartlett test result of 1,045.786 with $p < 0.001$. The factor loadings of each item on the scale are also more than the ideal value of 0.50.

Based on the above results, it can be said that the construct or structural validity of the scales were adequate. Hence, meaningful analysis and interpretations can be done on them.

### *Descriptive and Correlation Analyzes*

Pearson correlation coefficient was used to establish the kind of relationship that exists between the study variables. The results of the descriptive statistics and correlation analysis are displayed on Table 5.3 below.

Table 5.3: Descriptive and correlation results

| **Variable** | **Mean** | **Standard Deviation** | **Organizational Culture** | **Individual Creativity** |
|---|---|---|---|---|
| Organizational Culture | 3.67 | 0.730 | 1 | |
| Individual Creativity | 3.94 | 0.639 | $0.264^{**}$ | 1 |
| **Organizational Culture Dimensions** | | | | |
| Clan Culture | 3.75 | 0.770 | | $0.232^{**}$ |
| Adhocracy Culture | 3.61 | 0.851 | | $0.230^{*}$ |
| Market Culture | 3.67 | 0.831 | | $0.279^{**}$ |
| Hierarchy Culture | 3.64 | 0.755 | | $0.217^{*}$ |

Correlation is significant at 0.01 level (2-tailed). N = 124.

Table 5.3 above shows the average of the responses given to the statements in the organizational culture and individual creativity scales. When the table is examined, it is seen that Organizational and its sub-dimensions (clan, adhocracy, market and hierarchy culture) have means of 3.67 and 3.75, 3.61, 3.67 and 3.64 respectively. The average/mean of individual creativity is 3.94. This implies that the organizational culture of the participating firms and the creativity of their employees are above average on a five-point Likert scale. In addition, the relationship between the study variables is shown on the same table. It is observed that there is a statistically significant low positive relationship between organizational culture and individual creativity. It is also seen that there is a low positive and statistically significant relationship between the sub-dimensions of organizational culture and individual creativity.

## *The Model Regression: Linking Organizational Culture to Individual Creativity*

The regression analysis of the interaction in Equation (5.1) was conducted to determine the effect of organizational culture on individual creativity. Before the regression analysis, the following regression model was proposed.

Table 5.4: Regression results of the effect of organizational culture on individual creativity

| **Dependent Variable** | $R^2$ | **Independent Variable** | **B** | **Std. Error** | *t* | **F** | **P** |
|---|---|---|---|---|---|---|---|
| Individual Creativity | 0.07 | Constant | 3.095 | 0.286 | 10.836 | 9.019 | 0.003* |
| | | Organizational Culture | 0.231 | 0.076 | 3.019 | | |

$$\textit{Individual Creativity} = \beta_0 + \beta_1 \textit{(Organizational Culture)} + \epsilon \quad (5.1)$$

Where $b_0$ is the constant value and $\epsilon$ is the error term.

Given the estimates in Table 5.4, the obtained adjusted $R^2$ of 0.07 implies that at least around 7 % of the variances in individual creativity can be explained by organizational culture. Besides this, the "F-value" of 9.019, supported by the obtained P-value of 0.003 ($p < 0.05$) conspicuously reveal the model's overall significance, thus, implying that the outcomes can be reckoned with. These findings buttress the first hypothesis that "Organisational culture positively affects individual creativity." Accordingly, H1 was accepted based on the regression results. These findings coincide with a couple of other researchers in the literature (Gupta, 2011; Navaresse, Yauch, Goff, & Fonseca, 2014). After the regression analysis, we obtained the estimated model as seen in Equation 5.2;

$$\textit{Individual Creativity} = 3.095 + 0.231 \textit{(Organizational Culture)} + \epsilon \quad (5.2)$$

## The Effect of Sub-dimensions of Organizational Culture on Individual Creativity

The following regression analysis was conducted to examine the impact of sub-dimensions (clan, adhocracy, market and hierarchy culture) of organizational culture on individual creativity.

Table 5.5: Regression results of the effect of sub-dimensions of organizational culture on individual creativity

| **Dependent Variable** | **$R^2$** | **Independent Variable** | **B** | **Std. Error** | ***t*** | **F** | **P** |
|---|---|---|---|---|---|---|---|
| Individual Creativity | 0.054 | Constant | 3.220 | 0.280 | 11.509 | 6.918 | 0.010* |
| | | Clan Culture | 0192 | 0.073 | 2.630 | | |
| Individual Creativity | 0.053 | Constant | 3.318 | 0.245 | 13.546 | 6.837 | 0.010* |
| | | Adhocracy Culture | 0.173 | 0.066 | 2.615 | | |
| Individual Creativity | 0.078 | Constant | 3.156 | 0.251 | 12.577 | 10.300 | 0.002* |
| | | Market Culture | 0.214 | 0.067 | 3.209 | | |
| Individual Creativity | 0.047 | Constant | 3.273 | 0.278 | 11.771 | 6.010 | 0.016* |
| | | Hierarchy Culture | 0.183 | 0.075 | 2.452 | | |

According to the results of regression analysis displayed on Table 5.5, an $R^2$ (percentage of explained variance) and F (degree of significance of the regression) values of 0.054 and 6.918 show that individual creativity can be explained by clan culture and the regression result is statistically significant ($p < 0.050$). Based on this, the second hypothesis was accepted. Adhocracy culture regression analysis was found to be positively significant. F value of 6.837, t value of 2.615, B value of 0.173 and $p < 0.050$ indicate that the regression is significant. Accordingly, the third hypothesis was retained. In addition, according to the results of the regression analysis, an $R^2$ value of 0.078 and F value of 10.300 values show that market culture has a positive effect on individual creativity and the result is statistically significant ($p < 0.050$). Based on this, the fourth hypothesis was upheld. Finally, the hierarchy culture regression analysis was found to be positively significant. An F value of 6.010, t value of 2,452, B value of 0.183 and $p < 0.050$ indicate that the result is statistically significant. Based on the finding, the fifth hypothesis was also accepted.

## *Conclusions, Policy Implications, and Research Limitations*

This study was carried out in Büsan organized industrial site of Central Anatolia Province, Konya- Türkiye to determine the effect of organizational culture of packaging firms on the creativity of their employees. Overall, the results of the inferential analyzes showed a statistically significant positive effect of organizational culture on individual creativity. The correlation analysis also revealed a statistically significant positive relationship between the two variables. Furthermore, the results showed that the dimensions of organizational culture (Klan, Adhocracy, market and hierarchy culture) have significant positive effect on individual creativity. These results concur with findings of other researchers in the literature (Gupta, 2011; Navaresse et al., 2014; Ogbeibu et al., 2018). Based on the data analyzes and findings, all the hypotheses of the study were accepted.

Considering that the importance of SMEs cannot be overemphasized in an economy (Onifade et al. 2022), we believe that the outcomes have essential implications for policy makers (especially for SMEs) in their pursuit of creating congenial organizational culture to increase employees' creativity at workplace. Policy makers as well as managers of SMEs need to accentuate and promote cultures that have the propensity of positively affecting the attitudes, perceptions, actions and thoughts of individual employees towards creativity. They need to institute measures that discourage bad cultures that have negative consequences of derailing employees' creative efforts. Furthermore, it is important to workout the modalities of enabling workers have a good work life balance that can enhance more creative behaviors. In a nutshell, the current results emphasize the ramifications of good workplace culture on employee creativity.

It cannot be said that this study is free from certain limitations. Firstly, the study was restricted only to packaging firms in the Büsan Organized Industrial Zone operating in Konya. Therefore, it is unclear to what extent this study can extend its findings to other businesses across the country. For this reason, it is recommended that future researchers focus on businesses not included in the current research. Secondly, mediating variables were not included in the conceptual framework/model of the study, thus, other studies can add these factors to understand their potential roles.

## Editors Note

The examination of the effect of organizational culture on individual creativity within packaging firms in the Büsan Industrial Site of Central Anatolia Province, Konya-Türkiye, holds significant implications for sustainability within the packaging industry and beyond. Organizational culture plays a crucial role in shaping employee behaviors, attitudes, and interactions within the workplace. By investigating how different facets of organizational culture, such as innovation orientation, collaboration, and risk-taking, influence individual creativity, this research offers valuable insights into fostering a culture of innovation and sustainability within packaging firms. Enhanced individual creativity can lead to the development of novel packaging solutions, eco-friendly materials, and efficient production processes that contribute to environmental conservation and resource optimization. Moreover, a culture that values and nurtures creativity foster employee engagement, job satisfaction, and organizational commitment, leading to improved performance, competitiveness, and long-term sustainability for packaging firms operating in Büsan Industrial Site and beyond. Therefore, this empirical investigation has the potential to inform strategic interventions aimed at cultivating a supportive organizational culture conducive to creativity, innovation, and sustainability within the packaging industry in Konya-Türkiye.

## References

Agars, M. D., Kaufman, J. C., Deane, A., & Smith, B. (2012). Fostering individual creativity through organizational context: A review of recent research and recommendations for organizational leaders. In M. Mumford (Ed.), *Handbook of organizational creativity*. Boston: Elsevier/Academic Press.

Aktaş, E., Çiçek, I., & Kıyak, M. (2011). The effect of organizational culture on organizational efficiency: The moderating role of organizational environment and CEO values. *Procedia—Social and Behavioral Sciences, 24*, 1560–1573. <https://doi.org/10.1016/j.sbspro.2011.09.092>.

Alvesson, M. (2012). *Understanding organizational culture.* Cornwall: Sage.

Amabile, T. M. (2017). In pursuit of everyday creativity. *The Journal of Creative Behavior, 51*(4), 335–337.

Arif, M., Qaisar, N., & Kanwal, S. (2022). Factors affecting students' knowledge sharing over social media and individual creativity: An empirical investigation in Pakistan. *The International Journal of Management Education, 20*(1), 100598.

Bharadwaj, S., & Menon, A. (2000). Making innovation happen in organizations: Individual creativity mechanisms, organizational creativity mechanisms or both? *Journal of Product Innovation Management, 17*(6), 424–434.

Borg, I., Groenen, P. J., Jehn, K. A., Bilsky, W., & Schwartz, S. H. (2011). Embedding the organizational culture profile into Schwartz's theory of universals in values. *Journal of Personnel Psychology, 10*, 1–12.

Cameron, K. S., & Quinn, R. E. (1999). *Diagnosing and changing organizational culture.* USA: Addison Wesley Publishing Company.

Cameron, K. S., & Quinn, R. E. (2011). *Diagnosing and changing organizational culture: Based on the competing values framework* (3rd edn). San Francisco: Jossey-Bass.

Çekmecelioğlu, H. G., & Günsel, A. (2013). The effects of individual creativity and organizational climate on firm innovativeness. *Procedia—Social and Behavioral Sciences, 99*, 257–264. <https://doi.org/10.1016/j.sbspro.2013.10.493>.

Csikszentmihalyi, M., & Sawyer, K. (2014). Shifting the focus from individual to organizational creativity. Dordrecht: Springer. In M. Csikszentmihalyi (Ed.), *The systems model of creativity* (pp. 67–71). New York: Springer.

Damaraju, N. L., Barney, J. B., & Dess, G. G. (2021). Do stringent bankruptcy laws always deter entrepreneurial activities? A study of cultural influences. *Entrepreneurship Theory and Practice, 45*(2), 418–439.

Demir, C., Unnu, N. A. A., & Erturk, E. (2011). Diagnosing the organizational culture of a Turkish pharmaceutical company based on the competing values framework. *Journal of Business Economics and Management, 12*(1), 197–217.

Fan, H., Li, C., Xue, C., & Yu, M. (2023). Clan culture and patterns of industrial specialization in China. *Journal of Economic Behavior & Organization, 207*, 457–478.

Glăveanu, V. P., & Beghetto, R. A. (2021). Creative experience: A non-standard definition of creativity. *Creativity Research Journal, 33*(2), 75–80.

Grover, V., Tseng, S.-L., & Pu, W. (2022). A theoretical perspective on organizational culture and digitalization. *Information & Management, 59*(4), 103639.

Gül, A. Y. D. (2019). *An investigation of the effects of authentic leadership on individual creativity: A research within packaging sector.* Istanbul: Bahcesehir University.

Gumusluoglu, L., & Ilsev, A. (2009). Transformational leadership, creativity, and organizational innovation. *Journal of Business Research, 62*(4), 461–473. <https://doi.org/10.1016/j.jbusres.2007.07.032>.

Gupta, B. (2011). Organisational culture and creative behaviour: Moderating role of creative style preference. *International Journal of Innovation and Learning, 10*(4), 429–441.

Handy, C. B. (1996). *The gods of management: The changing work of organizations.* New York: Oxford University Press.

Harrison, R. (1995). *The collected papers of Roger Harrison.* London: McGraw-Hill.

Hofstede, G., Hofstede, G. J., & Minkov, M. (2010). *Cultures and organizations: Software of the mind-intercultural cooperation and its importance for survival.* London: McGraw-Hill.

İşcan, Ö. F., & Timuroğlu, M. K. (2007). Örgüt kültürünün iş tatmini üzerindeki etkisi ve bir uygulama. *Atatürk Üniversitesi İktisadi ve İdari Bilimler Dergisi, 21*(1), 119–135.

Low, W. W., Abdul-Rahman, H., & Zakaria, N. (2020). Organisational culture of Malaysian international construction organisations. *International Journal of Construction Management, 20*(2), 105–121.

Luthans, F., & Doh, J. P. (2017). *International management: Culture, strategy, and behavior* (10th edn). New York: McGraw-Hill.

Meng, H., Cheng, Z.-C., & Guo, T.-C. (2016). Positive team atmosphere mediates the impact of authentic leadership on subordinate creativity. *Social Behavior and Personality: An International Journal, 44*(3), 355–368. <https://doi.org/10.2224/sbp.2016.44.3.355>.

Muñoz-Doyague, M. F., & Nieto, M. (2012). Individual creativity performance and the quality of interpersonal relationships. *Industrial Management & Data Systems, 112*(1), 125–145.

Navaresse, D. O., Yauch, C. A., Goff, K., & Fonseca, D. J. (2014). Assessing the effects of organizational culture, rewards, and individual creativity on technical workgroup performance. *Creativity Research Journal, 26*(4), 439–455.

Ogbeibu, S., Senadjki, A., & Gaskin, J. (2018). The moderating effect of benevolence on the impact of organisational culture on employee creativity. *Journal of Business Research, 90*, 334–346.

Oldham, G. R. (2003). Stimulating and supporting creativity in organizations. In S. E. Jackson, M. A. Hitt, & A. S. DeNisi (Eds), *Managing knowledge for sustained competitive advantage* (pp. 243–273). San Francisco: Jossey-Bass.

Onifade, S. T., Hakan, A., & Savaş, Ç. (2022). Modeling the impacts of MSMEs' contributions to GDP and their constraints on unemployment: The case of African's most populous country. *Studies in Business and Economics, 17*(1), 154–170. <https://doi.org/10.2478/sbe-20220011>

Perry-Smith, J. E. (2006). Social yet creative: The role of social relationships in facilitating individual creativity. *Academy of Management Journal, 49*(1), 85–101.

Pheysey, D. C. (2002). *USA & Canada*. London: Routledge.

Robbins, S. P., & Judge, T. A. (2016). *Organizational behavior, global edition* (17th edn). Harlow, England: Pearson Education Limited.

Schein, E. H., & Schein, P. A. (2016). *Organizational culture and leadership* (5th edn). Hoboken, NJ: John Wiley & Sons, Inc.

Schneider, B., Ehrhart, M. G., & Macey, W. H. (2013). Organizational climate and culture. *Annual Review of Psychology, 64*, 361–388.

Shalley, C. E., & Gilson, L. L. (2004). What leaders need to know: A review of social and contextual factors that can foster or hinder creativity. *The Leadership Quarterly, 15*(1), 33–53. <https://doi.org/10.1016/j.leaqua.2003.12.004>.

Shrestha, N. (2021). Factor analysis as a tool for survey analysis. *American Journal of Applied Mathematics and Statistics, 9*(1), 4–11.

Taggar, S. (2002). Individual creativity and group ability to utilize individual creative resources: A multilevel model. *Academy of Management Journal, 45*(2), 315–330.

Telli, S. (2016). *Örgüt kültürü ve bireysel yaratıcılık: bir alan çalışması* (Yüksek Lisans). Kahramanmaraş Sütçü İmam Üniversitesi, Kahramanmaraş, Türkiye.

Tesluk, P. E., Farr, J. L., & Klein, S. R. (1997). Influences of organizational culture and climate on individual creativity. *The Journal of Creative Behavior, 31*(1), 27–41.

Valmohammadi, C., & Roshanzamir, S. (2015). The guidelines of improvement: Relations among organizational culture, TQM and performance. *International Journal of Production Economics, 164*, 167–178.

Walia, C. (2019). A dynamic definition of creativity. *Creativity Research Journal, 31*(3), 237–247.

Zhou, J., & George, J. M. (2001). When job dissatisfaction leads to creativity: Encouraging the expression of voice. *Academy of Management Journal, 44*(4), 682–696.

Zhou, J., & George, J. M. (2003). Awakening employee creativity: The role of leader emotional intelligence. *The Leadership Quarterly, 14*(4–5), 545–568. <https://doi.org/10.1016/s1048-9843(03)00051-1>.

PART II

# ICT, Energy and Financial Sustainability

Within the context of business, the concept of sustainability pertains to conducting business operations in a manner that does not engender adverse effects on the natural environment, local communities, or broader society. The concept of sustainability in business typically encompasses two primary domains, namely the environmental impact of business operations and the social impact of business activities. The objective of a sustainable business strategy is to generate a favorable influence on at least one of the aforementioned domains. In instances where companies neglect to accept accountability, adverse consequences may arise, such as environmental deterioration, inequity, and societal impropriety.

The financial industry wields significant influence in promoting and financing sustainability initiatives. This includes facilitating the exploration and advancement of renewable energy sources, as well as endorsing enterprises that adhere to equitable and sustainable labor standards. The concept of sustainable finance pertains to investment choices that incorporate considerations of environmental, social, and governance (ESG) factors in relation to a given economic undertaking or venture. Environmental factors encompass the implementation of measures to address the climate crisis, such as mitigation strategies, as well as the utilization of sustainable resources. Social factors encompass a range of considerations such as the protection of both human and animal rights, ensuring consumer protection, and implementing diverse hiring practices. Governance factors pertain to the practices of managing, handling employee relations, and compensating employees in both public and private organizations.

Moreover, Sustainable energy refers to energy sources that are inexhaustible and have the potential to remain viable indefinitely. Sustainable energy sources are characterized by their ability to meet energy demands without the need for renewal or replenishment, thereby eliminating the risk of depletion or spoilage. The utilization of sustainable energy sources is deemed as the optimal solution to meet our energy demands. In addition, sustainable energy sources are known to have negligible environmental impact, minimal contribution to climate change, and relatively low cost. Whilst there is an initial investment required for the development and implementation of sustainable energy infrastructure, the energy generated from these sources is typically devoid of any direct monetary cost.

Based on the findings from the section in ICT, energy, and financial sustainability, various techniques can be employed to enhance real-life situations and promote sustainable development. Integrating ICT solutions into energy management systems help improve efficiency and decrease resource use. Through the utilization of advanced technology such as smart meters, sensor networks, and data analytics, organizations can enhance energy utilization, detect inefficiencies, and execute specific conservation strategies. Additionally, ICT platforms can enable real-time monitoring and management of energy usage, allowing for proactive actions to reduce waste and minimize environmental impact. Furthermore, digital tools and platforms have the capacity to empower customers to make well-informed choices regarding energy usage, encourage changes in behavior, and provide incentives for adopting sustainable practices.

Moreover, incorporating financial sustainability concepts into energy projects can improve their feasibility and long-term resilience. This entails implementing funding structures that prioritize environmental, social, and governance (ESG) factors and align with sustainable development goals. For example, initiatives related to green financing can attract funds for projects involving renewable energy, upgrades in energy efficiency, and investments in clean technologies. Additionally, progressive financial tools like green bonds, carbon credits, and impact investing funds can incentivize the involvement of private companies in sustainable energy projects. Furthermore, by promoting collaborations among financial institutions, government agencies, and industry players, we can harness the combined power of these entities to create mutually beneficial outcomes, mitigate risks, and stimulate investments in sustainable energy infrastructure. By utilizing information and communication technology (ICT) tools to enhance energy management practices and incorporating financial sustainability principles into energy projects, organizations can improve real-world conditions, foster economic well-being, and promote environmental responsibility.

JENEBA JOY TUCKER AND DR. MURAD ABDURAHMAN BEIN

# 6 The Impact of Credit Risk on Bank's Performance after the World Financial Crises: Evidence from Commercial Bank in Sierra Leone

ABSTRACT
Banks help economies grow. Banks provide cash to institutions or customers, whether people or businesses, in hopes of a specified return. They connect savers and needy people. This research examines how credit risk affects Sierra Leonean banks. The study employed quantitative methods and secondary data. The Bank of Sierra Leone, Sierra Leone Statistical Service, and some selected banks provided annual reports for the data. The estimations were based on 2008–2019 balanced panel data from five Sierra Leonean banks. Ordinal, totally modified, and dynamic ordinal least square were used to solve this challenge. Return on Assets (ROA) was a banking performance indicator (in terms of profitability). The research measured credit risk using NPLR, TLR, and PLLR.

As a consequence of the outcome of the analysis, each of the credit risk measures (NPLR, TLR, and PLLR) were inversely but significantly related with banks' performance. All the factors used for this analysis suggest that when bank performance improves, they will decrease. Despite this, there was only a causation in one direction between NPLR and ROA, two-way causality between ROA and PLLR and a unidirectional relationship between PLLR and NPLR.

On the other hand, the research suggested that bank management adopt a comprehensive credit risk policy as a means of contributing to the modernization and improvement of rules in credit risk management.

Keywords: Return on assets, non-performing loans, total loans, total deposit, provision of loan loss and Sierra Leone

## Introduction

Banks offer financial services that increase economic activities, connect lenders to borrowers and boost a nation's economy. Most economies determine banking sector tenacity with effectiveness and productivity in mind (Haberly & Wójcik, 2022). According to Takyi and Naidoo (2022) most banks generate cash through the provision of credit, which is risky. At a significant credit risk, the license may be revoked or terminated, putting depositors and the bank's integrity at risk. Thus, banks appear to take excessive risks to survive in the harsh financial climate and many high-risk banks fail (Ozili, 2020). Defaulting on loans has created a problem in Sierra Leone's banking industry. According to a study conducted by the Central Bank of Sierra Leone in 2011, non-performing loans increased from 16.2 % in 2009 to 17.6 % in 2010 (bsl.gov.sl). The success of the banking industry (measured in terms of profitability) has increased dramatically over the years. The data produced by the central bank still don't clearly demonstrate the impact of creditworthiness on bank outcome in the country (Osei-Assibey & Asenso, 2015).

The banking sector in Sierra Leone has expanded significantly over the last decade, leading to an unprecedented level of credit risk. Yet, significant resources have been invested in mitigating domestic credit risk. This study aims to address a gap in the literature by analyzing the impact of credit risk on banks' performance. A number of outstanding evaluations of credit risk as well as banking efficiency globally, including profitability (Qi et al., 2023). Sierra Leone has few research on this. A very few Sierra Leone credit risk studies employed pre-Basel II data. Amidu and Hinson (2006) and Boahene et al. (2012), premised on 1998–2003 and 2005–2009 data, may be addressed. The banking industry has been affected by license revocation, and the economic crisis in Sierra Leone. This research examines the effect of credit risk on five commercial banks: Rokel Commercial Bank, Standard Charted Bank, Union Trust Bank, FI Bank, and Ecobank, utilizing information collected from 2010 to 2019.

Credit risk was a key factor in the demise of Procredit Bank Sierra Leone. Poor supervision, non-performing loans, and administrative procedures were major issues too (bsl.gov.sl). In most industrialized nations, credit risk is a growing worry, and big changes need intentional efforts to build a trustworthy framework in other countries and offer banks with a platform for successful and profitable banking. Credit risk is a crucial weapon for all financial institutions. Therefore, the primary objective of this study is to examine the role of credit risk in enhancing bank performance in Sierra Leone in the wake of the global financial crisis.

This research is important for a number of reasons. First, the findings will serve as a notice to banks in Sierra Leone that, despite positive returns, credit risk should still be taken into account. Second, eliminating bank failures and establishing economic stability in Sierra Leone are both possible outcomes of improved risk management in the banking sector. And lastly the findings of this research would be significantly helpful toward officials and bank management to embrace credit risk management's influence on banks' performance (in terms of profitability). This protects banks and customers from business credit risk. Good credit risk management in banks, especially newcomers, is also profitable.

## Literature Review and Hypothesis Development

### *Concepts of Risk Management*

Credit risk research began in the 1900s. Financial organizations experience credit-granting risk, money and capital market risk, and operational risk every day, according to research. These dangers need risk management in any bank management setting (Archison, 2018). A comprehensive empirical study found that several indications contribute to bank failures. They include deregulatory concerns, a lack of client information, political intervention, corruption, fraud, bad bank management, a highly competitive banking industry, and irresponsible lending. Reckless lending and poor bank management, especially credit risk assessment and management, are the main culprits, frequently forcing banks to acknowledge minimal value of non-performing loans owing to a steady rise in default rates (Archison, 2018).

Research also shows that most banks have adopted poor risk evaluation systems and credit assessment methods, which has led to their repeated failure (Archison, 2018). Most financial organizations don't employ good credit evaluation methodologies to anticipate borrowers' credit repayment. Focusing on credit risk, Sierra Leonean banks and other financial institutions still face a lot of it. This includes default, equity, securitization, and concentration risks. The equivalent default risk assesses the possibility that bank borrowers will default (i.e., the principal amount as well as the interest). If banks utilize equity instruments like stock options, share splits, equity purchases, and others, equity risk arises.

Risk securitization examines the pace and impact of financial arrangements entered into by banks and other financial institutions to shield themselves from the many hazards they face every day, whereas the risk of concentration is a more general notion that applies to bank customers' deposits. Concentration risk is predicated on the chance or degree at something many depositors appeared to decide to withdraw a large quantity of money from banks and other lending institutions (especially microfinance institutions) at once. Concentration risk is highlighted because banks employ depositor funds to lend to borrowers, acting only as financial intermediaries (Schaffner, 2023).

## Compliance with the Basel Accord

The 1990s banking industry led changes with the first Basel Accord, which set minimum capital requirements for global banks. The Basel Accords were created slowly in the 1980s. Member States started the BCBS in 1974 to discuss banking supervision concerns. BCBS promotes "financial soundness" via global banking regulation and supervisory expertise. BCBS then tested and maintained banks' capital and regulatory systems.

The Basel Consultative Group on Bank Regulation devised three criteria—Basel I, II, and III—to form the Basel Accords. Base l I focuses on minimum capital requirements, Basel II on capital adequacy, and Basel III on public transparency and corporate control. The research company advises banks on capital, market, and operational risk. The accords guarantee money-related companies have enough liquidity to handle unanticipated events.

The Basel Committee for Banking Supervision has widened the framework through which a bank should manage credit risk. One of the requirements is that their credit risk policies be regularly evaluated and changed to account for the state of the economy. Anything may be predestined if contemporary technologies and systems are fully implemented. The banking industry should often do a thorough evaluation of their creditor base. As a result, they may develop a more nuanced understanding of the customer they are serving (Basel Banking Oversight Committee, 1999). Although these initiatives may lessen the amount of credit risk that banks face, they cannot guarantee that a loan opportunity will arise. This is far greater, and it would increase the bank's profits.

## Credit Risk Management

The ability to quantify credit risk is essential to the effective management of credit risk. The most important challenge for financial institutions is figuring out how to accurately quantify credit risks and portfolio ratios. This is due to the fact that as credit risk ratios increase, the potential return on investment on credit investments as well as the amount of capital that is required also increase (Triki & Abid, 2023). Muhammad and Garba (2014) identified two key instruments that might be used to analyze or quantify credit risk. They include the initial rate, as well as the cost of any advanced loans.

Borrowing is risky. The bank's credit portfolio makes up half to three-quarters of its worth, making credit risk crucial. Banking and profitability choices have long included credit risk. Greuning and Bratan (2003) define loan risk as a delayed or non-payment that might affect a bank's liquidity and income. If neither side has changed the exchange agreement, these researchers face credit risk. Credit risk comes from contractual parties' uneasiness in credit management (Isenberg, Sazu & Jahan, 2022). Credit risk depends on contractual parties' capacity to perform. Brixton et al. (2023) defined credit risk as a possible change in customers' net income due to late or non-payment. The Basel Committee on Banking Supervision argues that a bank borrower or debtor who fails to meet their commitments under criteria stipulated will be most readily identified as credit risk. These researchers concluded from the definitions and implications that credit risk is a condition that may lead to financial consequences if unchecked.

## The Banking Sector in Sierra Leone

Sierra Leone, a low-income, post-war coastal West African country with 8.42 million people, relies on agriculture and minerals for foreign cash. The Bank of Sierra Leone became the country's central bank in 1963. Commercial banks rule the financial markets. Sierra Leone has 14 commercial banks, 17 local banks, 59 FSA, 13 MFIs (two deposit and 11 credit only), and 2 mobile money operators. Commercial banks have 99 % of financial reserves. Two state-owned, eleven multinational, and one local banks. Commercial banking sector reserves represent 99 % of financial system assets at Le 5,29 trillion. The 11 multinational banks own almost 60 % of industry assets (bsl.gov.sl).

The bank system's capital adequacy ratio (CAR) grew from 40.10 % in 2020 to 41.16 %, showing enough buffer for losses. Asset efficiency has grown in recent years, yet it remains problematic. NPLs increased from 12.67 % in 2020 to 15.23 % in 2021. Return on investment has also reduced, with profitability going from 6.11 % to 5.36 % and return on capital (ROE) from 27.61 % to 23.86 %. Due to an expanding banking system, additional deposit accounts at commercial banks have also increased (bsl.gov.sl).

Despite many foreign banks entering in the recent decade, the two state-owned financial firms (Rokel Commercial Bank RCB and Sierra Leone Commercial Bank SLCB) seem to remain the main players with 28.6 % of assets, 36.2 % of deposits, and 23.8 % of loans. RCB and SLBC had long-standing asset concerns in 2014, depleting their financial bases and prompting government change. This droves the machine wide NPL increase. BSL momentarily managed the two banks and created a framework to limit corporate lending. The government is considering restructuring these two banks. Such a move might increase bank risks and lower company borrowing rates, affecting the country's monetary sustainability (bsl.gov.sl).

## Bank Performance (Profitability)

The manager's perspective on risk may also be gleaned from the bank's profitability. When it comes to taking chances, profitable banks don't care whether they lose money. In a similar vein, inefficiently managed financial institutions incur greater levels of debt. Investors need a reliable method of gauging profits. Profitability is critical for the bank's stockholders since it reflects the return on their investment (Karolina, 2023). In addition, profitability rankings are used to assess the overall performance of businesses. A metric of profitability used by these analysts is the return on assets. For their part, the researchers claim that ROA is preferable than ROE since it has lower financial requirements and less related hazards (Bashir et al., 2023). When considering ROA as a tool for producing money, it is feasible to compare businesses both inside and between sectors. Therefore, Return on Assets is a Crucial Profitability Metric (Devinaga, 2010).

According to Devinaga (2010), the two main categories into which researchers looked for factors that may impact the bank's efficiency and profitability. In addition to internal factors, there are also external ones. Husni (2011) demonstrates that banks have control over the components that make up internal profitability. Hence, it is up to the bank to determine the extent to which such factors may be used. Each factor influences the bank's income and costs. Some research publications have classified similar factors into two groups. Both financial and non-financial variables may affect the way in which income is reported. Both the financial and non-financial elements, such as the rate at which new bank branches are opened, have a major impact on a bank's financial status and condition (Ha

& Nguyen, 2023). The factors listed below influence how much money a bank makes.

Devinaga (2010) separates bank income into interest and non-interest. Customer deposits, overdrafts, and industrial deposits generate fund interest. Fees and stock gains are non-interest income. Banks earn all interest. The corporation earns 80 % from interest (Vong et al., 2009). Devinaga (2010) stated banking regulations expect banks to maintain adequate liquidity for everyday activity including customer withdrawals. If banks can amass enough cash, this is the fastest way to get capital from other locations, he added. Liquidity sustains the bank. Devinaga (2010) finds that excess assets and low cash affect bank profitability. Profit requires liquidity management.

Diversified firms pay banks, and include creditor advancement. Lending more may help banks expand financially (Opoku, 2015). Due to the significant risk of liquidity and nonpayment, banks should be careful (Devinaga, 2010). The bank's guaranteed subprime lending precipitated the 2008 U.S. credit crisis. As real estate rates fall, they lend rapidly, unwilling to pay back the loan and interest, which contributes to bank failures and sub-loans to debtors or consumers (Opoku, 2015). Sierra Leonean banks gained throughout the recession despite mounting non-performing loans. IMF stress testing shows mismanaged bank reserves increase bad loans. Due to financial decline, the bank may fail. Hence, improving credit reduces monthly costs. Vong et al. observed in 2009 that bad debt harms Macao banks. Another credit indicator is loan-to-asset ratio (LOLA). Vong et al. (2009) discovered LOLA harms profitability, not development or outcomes. Vong (2005)'s predictions were confirmed. Vong et al. (2009) evaluate loan dependability using non-performing loans, debt guarantees, loan loss ratios, and advances.

## Credit Risk and Bank Performance

Banks lend money to people, corporations, and governments for interest on deposits and collateralized debt. Its three pillars are debt magnitude, default risk, and collection risk. Banks must regularly evaluate loan applications for credit risk reduction since they are generally immune to economic instability (Enad & Gerinda, 2022). Credit risk is primarily caused by constantly changing interest rates, small banks' bandwidth, unproductive lending guidelines, rapidly changing interest rates, weak institutions, low capital and liquidity levels, low-interest loan schemes, loan activities, loan performance ratings, policy interventions, and unsuitable monetary authority regulation. Credit risk is the danger of not being paid for credit commitments, according to Basel, BCBS (2001) and Gostineau Banking Supervisory Committee (1992). Credit operations include recession, nonpayment due to rejection/default or credit rating modifications, and Basel Supervisory Committee restructuring. Heffernan defines credit risk as the risk that investments or debts will never be recovered in the event of an impending monthly payment or potential repayment failures, while Banking Oversight-BCBS (1999) defines creditworthiness as a motivator not to satisfy its obligations with bank debtors or trade creditors, under availability to discuss. No matter what occurs, the company's solvency and securities value will drop. Credit risk is significant because a few major clients might produce catastrophic losses that could bankrupt a company (Cheng et al., 2020).

The banking business relies on data, yet debtor accountability causes disparities. Only main or secondary information is acceptable. Its insurance system includes two sorts (Godbillon-Camus & Godlewski, 2005). Risk cutoffs vary reliably on the scale. Negotiating Parties Bank bond costs may indicate data processing conditions (Tung, 2020). Financial collapses may disrupt growth patterns (Vermeulen et al., 2021). Companies should consider low-level hazards and pressure rises (Olivier Hassid, 2008). Credit risk management is to optimize projected returns at a given risk level, restrict credit risk within an acceptable range, and consider credit risk's influence on banks' bottom lines (Kargi, 2011). BCBS warns of bank funding risk (1999). Contractual risk is lower than credit risk in investment products, interbank loans, and currency trading. Possible pre-contractual incompetence. Brownbridge (1998) claims Internal Credit supports most failed banks. After write-downs, nonperforming loans hurt a bank's investment portfolio even more.

Credit is needed when a consumer's cash flow is inadequate (Hiller & Jones, 2022). Every debt threatens earnings, and "individual" reimbursements or loan payments were never made before expiration. This risk is critical in banking (Leo & Maddulety, 2019). Volatility—changes in advantage or loss—has always caused the danger. Income volatility determines variance, whereas risk significance determines negative results. The chance of a disaster depends on its scale and gravity. Yet, the bank's default loss wouldn't need high-risk. Securities, financing, asset securitization, Basel Agreement compliance, and the Bureau of Financing Regulation help banks decrease losses and promote safety (Kolapo, Funso, Ayeni, Oke & Ojo, 2012) Credit instruments allow banks to lend more at lower interest rates. Koehn and Santomero (1980), Kim and Santomero (1988), and Athanasoglou et al. have proven that bank risks improve profitability and security over time (2005). According to Do and Phung (2020), a bank's profitability may affect its capacity to predict, avert, or monitor hazards that may reduce losses. The bank's loan ratio will rise, but its benefit will fall. Senior management is well-versed in this. In the economic crisis, especially in a competitive environment, managers must be inspired to follow. They suggest that even conservative managers may struggle to overcome the market's greed. There is a direct correlation between high risk and big potential reward in investing. It's easy to notice whether someone is financially finicky or the subject of unfavorable rumors. According to research by Lee et al. (2019), certain financial institutions engage in bad practices despite having access to a reliable lender. As borrowers withdraw money, the bank's reserves dwindle and the institution may be forced to close if it cannot get emergency funding.

The role of the bank in the economy is as crucial as ever. A prosperous banking sector may benefit the economy as a whole because it is better able to weather unexpected downturns and promote growth across the financial sector (Winful & Irenee, 2020). Also, the academic research and administration of the bank drew scrutiny from the aspects impacting the performance of the bank. Internal factors are evaluated using tools like capital size, credit risk management, and cost accounting. The banking industry, like many others in the financial sector, requires regular risk assessments. Poor asset quality along with inadequate liquidity are the dynamic capability of financial meltdown and constitute the central aspect of credit threat and unpredictability, as well as the effect on bank income has been researched with significant interest from researchers.

Many empirical data and studies show how credit risk affects bank profitability. Saeed and Zahid (2016) evaluated UK corporate profitability and risks from 2007 to 2015. Recent research suggests NPL reduces financial crises and bank intensity (Ozili, 2017). The initial loan differs from studies on mutually advantageous relationships (Messai & Jouini, 2013) in Italy, Greece, and Spain between 2004 and 2008. Despite Swedish banks' inversely proportional loan-to-capital ratios, Hosna, Manzura, and Juanjuan (2009) discover that banks' adverse ties vary. Expert ratios estimate credit risk. The loan loss ratio relative to gross loans measures bank asset quality. Higher ratios indicate safer and more efficient lending portfolios. Interest payments are another indicator of a loan's quality. Studies across countries showed legislative differences (Houcine & Houcine, 2020).

Brewer (1989) uses the bank-asset ratio, or the ratio of banks' loans to their total assets, to assess financial institution risk (LTAs). LTA and risk management may be linked since bank loans are deflationary and unstable. If credit risk management practices change, the LTA may have no effect on banks' risk behavior (Mohd Amin & Abdul-Rahman, 2020). Bourke emphasizes how credit risk affects business (1989). Profits are low because banks favor high-risk loans. Commercial banks lose less interest (Pham & Nguyen, 2023). Felix and Claudine (2008) found that financial institution TLs and NPLs are linked to ROE owners' returns and commodity profitability. Kargi (2011) examines credit risk and Nigerian bank profitability. According to research, credit risk management affects Nigerian bank profitability. The bank's liquidity and suffering are threatened by the number of loans and the growth of bad loans and deposits. Risk is anything that might harm one's finances, income, or production now or in the future (Mbeba 2007). Indiael Kaaya and Dickson Pastory (2013) examined credit risk and financial performance using return on assets. The credit risk index was negatively correlated with bank performance, suggesting that higher credit risk leads to worse bank performance. Overprotection reduces bank profitability. This study found that aggressive managers hurt Tanzanian banks' earnings.

Both developed and developing country research shows positive and negative connections between credit risk and bank results. Financial efficiency is strongly correlated with benefit, but weakly with resource appropriateness, asset quality, dependability, and liquidity, according to Ogilo (2012). Ethiopian banks' profitability may suffer (Kapur & Gualu, 2011). When measuring asset quality and efficiency, traditional Pakistani banks' NPA ratios hurt profitability (Akhtar, Ali, & Sadaqat, 2011). Indian private banks have poorer asset efficiency and profitability (Khalid & Amjad, 2012). Notwithstanding worries regarding the relationship between distressed lending and rural bank performance, Ghanian banks made a profit between 2006 and 2010 (Afriyie & Akotey, 2013). This research uses Ghanian data to analyze how credit risk management affects bank profitability and non-performing loans (Boahene, Dasah, & Agyei, 2012). They said NPLs boosted financial performance. Due to high tax and fee commitments. Ogboi and Unuafe (2013) found that competent capital fund credit risk management increases Nigerian banks' financial output. A 15-year research found that NPL-bank collaboration boosts credit growth (Onyekachi & Okoye, 2013). Hence, interest income indicates bank efficiency.

Athanasoglou, Brissimis, and Delis (2005) used panel data analysis to study how credit risk affects Greek financial services industry profitability. They found that credit risk hurts profitability. Credit risk lowers profitability. Felix and Claudine (2008) examined how credit risk management affects corporate operations. The bank's success is correlated with the financial firm's non-performing loan ratio. In a paper on effective loan management, Aremu, Suberu, and Oke (2010) suggested non-performing lending as a great option for Nigerian banking industry assets that don't operate. Nigeria's survival is threatened. Alalade et al. (2014) examine Lagos bank profitability and credit risk management. The research examined its premise to discover how credit risk affects bank profitability and credit risk. This research examined a bank's credit risk management's profitability-boosting potential. Using standardized questionnaires, respondents provided data for analysis. The correlation coefficient determined whether credit risk assessment affected profitability. Credit risk diminishes benefit, hence credit risk management in Lagos is important.

Credit risk affected Ethiopian commercial banks, according to Gizaw et al. (2013). Eight commercial bank surveys over 12 years provided the additional data. STATA version 11 processes extensive statistics and panel regression data to calculate credit risk factors. Ethiopian commercial banks suffer from non-compliant loans. Hence, researchers suggested that Ethiopian commercial banks strengthen their credit risk management method to improve their finances. Loans are currently the biggest credit risk, according to the Basel Council for Banking Supervision (1999). Both international banks should also consider how to monitor, manage, and mitigate credit risk. This guarantees the Bank has enough reserves to cover all expenditures and be reimbursed. Basel II analyzes banks' funding needs to protect them from financial and operational risks.

According to Metawa et al. (2023) credit risk's impact on bank outcomes. Methodological results differ. The observational study' logistic regression panels are too few to adequately affect the bank's credit risk. Finally, the research on the credit risk impact on the banking industry in Sierra Leone is dispersed (Davies, 2023). To understand how credit risk affects banks, particularly Sierra Leone, additional study is needed to solve the aforementioned issues. Based on the foregoing arguments, we formulate the following hypothesis:

$H_{1:}$ Credit risk has impact on banks performance in Sierra Leone

$H_2$: Non-performing loans has impact on banks performance in Sierra Leone

$H_3$: Total loans on total deposit have impact on banks performance in Sierra Leone

$H_4$: Impaired loans ratios have impact on banks performance in Sierra Leone

## Methodology

The aims of the present investigation are met via the use of quantitative analytic techniques. When quantitative methods are used, results may be trusted since there is more room for analysis in the review. Data from quantitative studies are reliable because they may be generalized to include both the study's intended population and people in similar circumstances who were not included in the study (Rahman et al., 2022).

The current sample is drawn from the banking sector in Sierra Leone's economy. There are now 14 recognized commercial banks in Sierra Leone, providing a wide variety of banking and financial services to the country's residents. There are a total of 14 banks, 10 of which are owned and run by non-natives (bsl.gov.sl).

This study used improbable sampling, specifically purposive sampling. Sierra Leone has 14 regulated commercial banks. The research profiles Sierra Leone's top five commercial and universal banks. It utilizes the five main Sierra Leone stock exchange banks. The Sierra Leone stock market has few banks; hence the research utilizes purposeful sampling to reach the necessary sample size of five. This collection includes historic and new banks. Hence, the researcher's test results were improved.

Where applicable to this research, data for evaluation was produced or received from various banks' Annual Audited Accounts available on the official websites of the chosen institutions. Several researchers, notably in the field of finance, drew on this secondary data source and economic sector (Asare et al., 2015). The study also makes use of information gleaned from journals and newsletters. Audited yearly reports from Rokel Commercial Bank, Standard Charted Bank, Union Trust Bank, FI bank, and Ecobank were obtained. Time series patterns were measured using a knowledge score for up to 12 years for regression and correlation analyzes. The study's specific time frame was 2008–2019.

This study used a Panel Data Analysis model to assess the relationship between credit risk and bank performance at a sample of Sierra Leonean institutions. Time-series data, such as that provided by panel models, looks back over a period during which the entity's behavior has been repeatedly observed (Xu et al., 2022). Experts may use this method to keep an eye on things like social aspects, gaps in strategy approaches across firms, or things that haven't changed yet through time via factors that can't be measured (national laws, government directives and global understanding). It's useful for reflecting individual differences among people. It also offers ever-increasing insight, freedom, discretion, fewer interdependent factors, and success (Gujarati & Sangeetha, 2007). To accomplish its goals, this research used a multivariate statistical analytic framework. They included the causality test, dynamic ordinary least squares, unit root test analysis, cointegration regression, and completely modified ordinary least squares (FMOLS).

## Study Model

$$\text{ROA} = \frac{\textit{NET INCOME}}{\text{TOTAL ASSETS}}$$

$$\text{NPLR} = \frac{\textit{NON - PERFORMING LOANS}}{\textit{TOTAL LOANS}}$$

$$\text{TLR} = \frac{\textit{TOTAL LOANS}}{\textit{TOTAL DEPOSITS}}$$

$$\text{PLLR} = \frac{\textit{PROVISION OF LOAN LOSS}}{\textit{NON - PERFORMING LOANS}}$$

$$\text{LR} = \frac{\textit{TOTAL EQUITY}}{\textit{TOTAL LIABILITIES}}$$

Hence, based on the above, the research suggests the following model to assess how credit risk affects bank performance after the global financial crisis. 5 commercial banks in Sierra Leone's evidence.

$$\text{ROA} = f(\text{NPLR}, \text{TLR}, \text{PLLR}) \tag{6.1}$$

$$\text{ROA} = \alpha_0 + \beta_{1i}\,\text{NPLR}_{it} + \beta_{2i}\,\text{TLR}_{it} + \beta_{3i}\,\text{PLLR}_{it} + \varepsilon_{it} \tag{6.2}$$

$$\text{NPLR} = f(\text{ROA}, \text{TLR}, \text{PLLR}, \text{BS}, \text{FL}) \tag{6.3}$$

$$\text{NPLR} = \alpha_0 + \beta_{1i}\,\text{ROA}_{it} + \beta_{2i}\,\text{TLR}_{it} + \beta_{3i}\,\text{PLLR}_{it} + \beta_{4i}\,\text{BS}_{it} + \beta_{5i}\,\text{FL}_{it} + \varepsilon_{it} \tag{6.4}$$

Where;

$\alpha_o$ = constant

β1–β2–β3–β4–β5 = coefficients

= coefficients

i = 1 … n

t = time period

$NPLR_{it}$ = the ratio of non-performing loans to total loans

$TLR_{it}$ = the ratio of total loans to total deposit

$PLLR_{it}$ = the ratio of provision of loan loss to non-performing loans

LE = Leverage

BS = Log of total net asset

$\varepsilon_{it}$ = error term

This balanced regression panel analysis study is objective. E-views (9) summarizes dependent and independent factors.

## Results

Initially, the research estimates descriptive data from Table 6.1 on credit risk's impact on Sierra Leone's banking industry using return on asset as the dependent variable and non-performing loans, total loans, and provision of loan loss as independent factors.

According to the research, ROA earns 2.96 million Leones per year, 2.80 million, and 7.00 million. NPLR had 10.58 million Leones annually as a mean, 8.35 million as median, and 36.1 as highest. Again, PLLR had a mean, median, and maximum of 9.66, 7.97, and 48.90 million Leones each year. TLR's mean, median, and maximum are 42.15, 36.6, and 97.65 million Leones per year, respectively. Bank size (BS) averages 29.25 million Leones every year, medians 25.45 million, and maxes out at 64.44 million. Finally, leverage (LV) averages 34.42 million Leones each year, medians 30.32 million, and maxes out at 81.72 million.

Table 6.1: Descriptive statistics

| | ROA | NPLR | PLLR | TLR | BS | LE |
|---|---|---|---|---|---|---|
| Mean | 2.960033 | 10.58130 | 9.659100 | 42.15170 | 29.25493 | 34.42169 |
| Median | 2.795000 | 8.350000 | 7.965000 | 36.59900 | 25.45275 | 30.32010 |
| Maximum | 7.000000 | 36.10000 | 48.88500 | 97.64600 | 64.44300 | 81.71830 |
| Minimum | 0.070000 | −0.197000 | 0.000000 | 1.459000 | 4.436000 | 3.543500 |
| Std. Dev. | 1.489376 | 8.193736 | 10.49437 | 21.82650 | 13.99035 | 17.86879 |
| Skewness | 0.536746 | 0.724475 | 2.190740 | 0.771564 | 0.636332 | 0.710621 |
| Kurtosis | 3.128105 | 3.046124 | 7.570532 | 3.124431 | 2.737113 | 2.983260 |
| Jarque-Bera | 2.921989 | 5.253966 | 100.2178 | 5.991810 | 4.221958 | 5.050517 |
| Probability | 0.032005 | 0.072296 | 0.000000 | 0.049991 | 0.121119 | 0.080038 |
| Sum | 177.6020 | 634.8780 | 579.5460 | 2,529.102 | 1,755.296 | 2,065.302 |
| Sum Sq. Dev. | 130.8763 | 3,961.101 | 6,497.776 | 2,8107.38 | 11,548.06 | 18,838.33 |
| Observations | 60 | 60 | 60 | 60 | 60 | 60 |

Source: Authors' computations.

## Correlation Matrix

The research then examined variable correlation. From Table 6.2, ROA shows a negative and significant association with NPLR and PLLR at 1 %, but a positive correlation with TLR at 5 %. NPLR has a negative and substantial connection with both PLLR and TLR at 1 % and 10 % significance. At 5 %, PLLR adversely impacts TLR. After finding association, the investigation determines stationarity.

## Unit Root Test Results

Following Abosedra et al. (2016), the unit root test determined variable stationarity. ADF-Fisher Chi Square (1981), based on the individual statistical test, and the Philip-Perron (PP) test were used to determine variable

Table 6.2: Correlation matrix

| | ROA | NPLR | PLLR | TLR | BS | LE |
|---|---|---|---|---|---|---|
| ROA | 1.000000 | | | | | |
| Prob | — | | | | | |
| NPLR | 0.372597*** | 1.000000 | | | | |
| Prob | (0.0034) | — | | | | |
| PLLR | −0.386545*** | −0.361055*** | 1.000000 | | | |
| Prob | (0.0023) | (0.0046) | — | | | |
| TLR | 0.198595** | −0.089689* | −0.160962** | 1.000000 | | |
| Prob | (0.0282) | (0.0956) | (0.0192) | — | | |
| BS | −0.360628*** | −0.188266 | 0.648744*** | 0.645131*** | 1.000000 | |
| Prob | (0.0046) | (0.1497) | (0.0000) | (0.0000) | — | |
| LE | −0.412095*** | −0.233341* | 0.600869*** | 0.687855*** | 0.997102*** | 1.000000 |
| Prob | (0.0011) | (0.0728) | (0.0000) | (0.0000) | (0.0000) | — |

Source: Authors' computations. Note: ***, ** and * are 1 %, 5 % and 10 % significant level respectively.

stationarity (1998). Both methods used constants and trends. These unit root test methods are better and can handle panel heterogeneities in group unit roots, notably in the ADF. Table 6.3 shows that all variables became stationary at first level in ADF but not PP.

## Cointegration Test Results

When the variables were stationary, the likelihood of long-term balance linkages had to be determined. Co-integration was used to analyze this incentive. To determine cointegration, Pedroni (1995, 1999) and Kao cointegration tests were utilized. The Pedroni test yields seven conclusions from four panel data (panel v-states, panel rho-stats, panel PP comments, panel ADF states) and group test results (Group PP-stats, Group ADF-stats, Group rho-stats). These approaches are crucial for testing panel and group statistics co-integration because they assume

Table 6.3: Unit root test

| | ADF | | | | PP | | | |
|---|---|---|---|---|---|---|---|---|
| | At Level | | At 1st Level | | At Level | | At 1st Level | |
| **Variables** | $\pi\tau$ | $\pi\vartheta$ | $\pi\tau$ | $\pi\vartheta$ | $\pi\tau$ | $\pi\vartheta$ | $\pi\tau$ | $\pi\vartheta$ |
| **ROA** | 0.0003 | 0.0026 | 0.0000*** | 0.0002*** | 0.0003 | 0.0001 | 0.0000*** | 0.0005*** |
| **NPLR** | 0.3015 | 0.0003 | 0.0001*** | 0.0001*** | 0.6654 | 0.0003 | 0.0000*** | 0.0001*** |
| **PLLR** | 0.0000 | 0.0001 | 0.0000*** | 0.0001*** | 0.0042 | 0.0239 | 0.0003*** | 0.0010** |
| **TLR** | 0.0594 | 0.0139 | 0.0044*** | 0.0305*** | 0.0352 | 0.0082 | 0.0003*** | 0.0015** |

Note: ***, ** and * are 1 %, 5 % and 10 % significant level respectively thus, $\pi\tau$ is with constant, $\pi\vartheta$ is with constant and trend.

Table 6.4: Pedroni cointegration test technique

| Alternative Hypothesis: Common AR Coefs. (Within-Dimension) | | | | |
|---|---|---|---|---|
| | **Statistics** | **Prob** | **Weighted Stats** | **Prob** |
| Panel v-Statistic | −1.307658 | (0.9045) | −1.049426 | (0.8530) |
| Panel rho-Statistic | 1.020935 | (0.8464) | 0.450064 | (0.6737) |
| Panel PP-Statistic | −3.783053*** | (0.0001) | −7.315291*** | (0.0000) |
| Panel ADF-Statistic | 1.353150** | (0.0120) | −1.905273** | (0.0284) |
| Alternative hypothesis: individual AR coefs. (between-dimension) | | | | |
| Group rho-Statistic | 1.767314 | (0.9614) | | |
| Group PP-Statistic | −4.398334*** | (0.0000) | | |
| Group ADF-Statistic | −9.033138*** | (0.0000) | | |

Source: Authors' computations. Note: ***, ** and * are 1 %, 5 % and 10 % significant level respectively.

conventional self-reliance mechanisms and residuals of the co-integration balance. Newey-automated West's bandwidth collection and Bartlet Kenel's cointegration estimate alternative assumed deterministic intercept and trend. The Kao cointegration test ensures serial linkages of the model error term with regressors. Series correlation errors do not violate the concept of erogeneity, which makes the normal least square estimator non-biased and correct. Only homoscedasticity and heteroscedasticity stable norm mistakes are invalidated and may misread and assume. The Kao cointegration, which yields the HAC, will fix this difficulty and explain the outcomes.

Tables 6.4 and 6.5 show that the components are co-integrated and balance long-term. In the Pedroni co-integration approach, eleven test results are used to verify variable co-integration if six or more are relevant. Hence, the null hypothesis with no co-integration was acceptable and the alternative hypothesis was accepted at 1%. The Pedroni test demonstrates that the PP statistics cointegrated at 5.47 (0.0000) and the ADF statistics at 1.35 (0.0120). PP co-integration was 4.39 (0.0000) in person and 9.03 across measures (0.0000). Kao cointegration calculates HAC variance (0.80).

Table 6.5: Kao cointegration test

| Variable | Coefficient | Std. Error | t-Statistic | Prob. |
|---|---|---|---|---|
| | | | −2.253372** | (0.0121) |
| Residual variance | | | 1.152470 | |
| HAC variance | | | 0.798101 | |
| RESID(-1) | −0.504028 | 0.103341 | −4.877311 | 0.0000 |
| D(RESID(-1)) | 0.092785 | 0.095882 | 0.967700 | 0.3380 |
| $R^2$ | 0.320670 | ADJ $R^2$ | 0.306518 | |

Source: Authors' computations. Note: ***, ** and * are 1 %, 5 % and 10 % significant level respectively.

## Long-Run Relationship

When the cointegration test showed that the variables were cointegrated, the next step was to determine their long-term associations. Phillips and Hansen's OLS, FMOLS, and DOLS were used (1990). From Tables 6.6, 6.7, and 6.8, a percentage shift in ROA decreases NPLR by 11.16 % in OLS, 4.24 % in FMOLS, and 3.83 % in DOLS. This confirms Saeed and Zahid (2016) and Ozili (2017). Hence, non-performing loans hurt Sierra Leonean banks.

In addition, the results of the FMOLS show that a percentage change in ROA results in a 4.66 % decline in PLLR and a 10.7 % drop in DOLS. The results imply that the PLLR will fall if the return on assets improves, which once again validates the conclusions of Alalade et al. (2014), Zahid (2016), and Ozili (2017).

In addition, the study demonstrates that a 1 % change in ROA results in a 3.6 % drop in TLR in OLS, 2.36 % decrease in FMOLS, and 4.15 % decrease in DOLS accordingly. These results show that when ROA for the banks in Sierra Leone grows, their TLR then falls, which contradicts the findings of Ozili, (2017), Alalade et al. (2014), and Suberu and Oke

Table 6.6: Ordinal Least Square (OLS)

| Variable | Coefficient | Std. Error | t-Statistic | Prob. |
|---|---|---|---|---|
| NPLR | −0.111606*** | 0.019586 | −5.698184 | (0.0000) |
| PLLR | −0.065780 | 0.016161 | −0.407059 | (0.6855) |
| TLR | −0.035926** | 0.006118 | −5.871936 | (0.0000) |
| R-squared | 0.342331 | | Mean dependent var | 2.960033 |
| Adjusted R-squared | 0.308728 | | S.D. dependent var | 1.489376 |
| S.E. of regression | 1.482862 | | Akaike info criterion | 3.674532 |
| Sum squared resid | 125.3362 | | Schwarz criterion | 3.779249 |
| Log likelihood | −107.2360 | | Hannan-Quinn criter. | 3.715493 |
| Durbin-Watson stat | 0.726944 | | | |

Note: ***, ** and * are 1 %, 5 % and 10 % significant level respectively.

Table 6.7: Fully Modified Least Square (FMOLS)

| Variable | Coefficient | Std. Error | t-Statistic | Prob. |
|---|---|---|---|---|
| NPLR | −0.042463*** | 0.025439 | −1.669215 | (0.0017) |
| PLLR | −0.046559* | 0.024666 | −1.887555 | (0.0653) |
| TLR | −0.023580* | 0.016141 | −1.460801 | (0.0507) |
| R-squared | 0.446766 | | Mean dependent var | 2.820764 |
| Adjusted R-squared | 0.364369 | | S.D. dependent var | 1.460832 |
| S.E. of regression | 1.164669 | | Sum squared resid | 63.75337 |
| Long-run variance | 1.269285 | | | |

Note: ***, ** and * are 1 %, 5 % and 10 % significant level respectively.

Table 6.8: Dynamic Ordinals Least Square (DOLS)

| Variable | Coefficient | Std. Error | t-Statistic | Prob. |
|---|---|---|---|---|
| NPLR | −0.038300*** | 0.002010 | −1.905204 | (0.0054) |
| PLLR | −0.106938*** | 0.000888 | −120.4650 | (0.0000) |
| TLR | −0.041485*** | 0.000896 | −46.30062 | (0.0000) |
| R-squared | 0.999964 | | **Mean dependent var** | 3.485767 |
| Adjusted R-squared | 0.999824 | | **S.D. dependent var** | 1.409108 |
| S.E. of regression | 0.018704 | | **Sum squared resid** | 0.002099 |
| Long-run variance | 3.87E-05 | | | |

Note: ***, ** and * are 1 %, 5 % and 10 % significant level respectively.

(2010), which say that ROA increases TLR. These researchers found that ROA increases TLR.

The research examines the association between NPLR and the bank-specific variable in Table 6.9. Non-performing loans negatively affect return on asset at 1 % significance. Indicating that NPLR % increases will lower ROA considerably. The country's banking industry is suffering from unrecovered customer debts. The link between PLLR and NPLR is positive but not statistically significant. It suggests that over time, there is no link between PLLR and NPLR. Moreover, there is a slight but favorable correlation between NPLR and TLR. This suggests that over the long term, there is no meaningful relationship between the two variables.

Nonetheless, there is a significant positive correlation between NPLR and BS. This demonstrates that over time, a change in NPLR by a certain amount reduces BS. The result demonstrates how the country's banks' sizes are affected by the pace at which the NPLR rises. So, it is crucial to have policies that can aid in recouping all of the loans given to consumers.

Last but not least, there is a long-term positive significant link between NPLR and LR, showing that a change in NPLR by % considerably raises LE.

Table 6.9: Using OLS to determine the relationship between NPLR and the bank-specific variables

| Variable | Coefficient | Std. Error | t-Statistic | Prob. |
|---|---|---|---|---|
| ROA | −3.251715*** | 0.570658 | −5.698184 | (0.0000) |
| PLLR | 0.034812 | 0.041832 | 0.115031 | (0.9088) |
| TLR | 0.118651 | 0.087324 | 0.213589 | (0.8316) |
| BS | −2.289496*** | 0.490520 | −4.667486 | (0.0001) |
| LE | 1.213000** | 0.433403 | 2.798781 | (0.0105) |
| R-squared | 0.378094 | | Mean dependent var | 10.58130 |
| Adjusted R-squared | 0.345747 | | S.D. dependent var | 8.193736 |
| S.E. of regression | 8.004124 | | Akaike info criterion | 7.046498 |
| Sum squared resid | 3651.762 | | Schwarz criterion | 7.151215 |
| Log likelihood | −208.3949 | | Hannan-Quinn criter. | 7.087458 |
| Durbin-Watson stat | 0.855109 | | | |
| | | | | |

Note: ***, ** and * are 1 %, 5 % and 10 % significant level respectively.

## Causality Test Result

It was crucial to establish the direction of causality between the variables under investigation since correlation does not always imply causation. The pairwise ganger homogeneous causality approach was then used in the research to determine the direction of causality between the variables. Table 6.10 contains the causality test's findings. The research revealed a one-way causal relationship between NPLR and ROA that was

Table 6.10: Pairwise Ganger causality test

| Null Hypothesis | Obs | F-Statistic | Prob. |
|---|---|---|---|
| NPLR→ROA | 50 | 2.54249* | (0.0899) |
| ROA→NPLR | | 1.58461 | (0.2163) |
| PLLR→ROA | 50 | 3.33002** | (0.0448) |
| ROA→PLLR | | 6.07269*** | (0.0046) |
| TLR→ROE | 50 | 0.05371 | (0.9478) |
| ROA→TLR | | 0.12618 | (0.8818) |
| PLLR→NPLR | 50 | 2.69667* | (0.0783) |
| NPLR→PLLR | | 0.85945 | (0.4302) |
| TLR→NPLR | 50 | 0.60776 | (0.5490) |
| NPLR→TLR | | 1.77298 | (0.1815) |
| TLR→PLLR | 50 | 0.80727 | (0.4524) |
| PLLR→TLR | | 1.09430 | (0.3435) |

Note: ***, ** and * are 1 %, 5 % and 10 % significant level respectively, → represent "does not Granger Cause."

consistent with (Su et al., 2021). The result suggests that NPLR affects ROA. This outcome is not unexpected given that non-performing loans have an impact on a bank's overall assets. The table once again demonstrates the two-way causal relationship between ROA and PLLR that is related to Shahbaz (2009) and Shahbaz and Islam (2011). Ultimately, PLLR and NPLR were shown to have a unidirectional connection.

## Discussion

This study uses audited annual reports from Rokel Commercial Bank, Standard Charted Bank, Union Trust Bank, FI Bank, and Ecobank in Sierra Leone. This research measured time series trends using a 12-year knowledge score for regression and correlation analyzes. The investigation spanned 12 years from 2008 to 2019.

Correlation was tested using the correlation test. From Table 6.2, ROA shows a negative and significant association with NPLR and PLLR at 1 %, but a positive correlation with TLR at 5 %. NPLR has a negative and substantial connection with PLLR and TLR at 1 % and 10 %, respectively. At 5 %, PLLR negatively correlates with TLR. Table 6.3 shows that no variable was stationary at level in both ADF and PP, but all became stationary at first level. After determining that the variables were stationary, two cointegration tests showed that they were cointegrated.

The OLS, FMOLS, and DOLS were used to determine the variables' long-term relationships once the variables' cointegration was established. Using all three approaches, it was determined that NPLR had a long-term, negative impact on ROA, reducing it by 11.16 % in OLS, 4.24 % in FMOLS, and 3.83 % in DOLS for every 1 % change in ROA. These results support the conclusions made by Saeed and Zahid in 2016 and by Ozili in 2017. So, it means that non-performing loans have a bad impact on Sierra Leone's banks. This result suggests that either the loans given by banks in Sierra Leone are weak or better strategies to reclaim them are not implemented. Nonetheless, it was shown that PLLR had a weakly significant negative correlation with ROA. The results of this study support those of Alalade et al. (2014), Zahid (2016), and Ozili (2017). Since bank loans are not well-backed, authorities have increased provisions, which is excellent news for the banking industry. Moreover, TLR has a long-term, strong unfavorable connection with ROA. These results contradict the conclusions of Ozili, (2017), Alalade et al. (2014), and Suberu and Oke (2010), which claim that ROA raises TLR. When ROA for the banks in Sierra Leone grows, their TLR then falls.

By include further factors like bank size and leverage, the second model was able to determine the link between NPLR and other variables. According to the analysis, there is a 1 % significant negative significant relationship between non-performing loans and return on assets. demonstrating that a percentage rise in NPLR would considerably reduce ROA. As the loans given to consumers are not being repaid, this does not bode well for the nation's financial industry and is adversely harming its performance. The link between PLLR and NPLR is favorable but not statistically significant. It suggests that PLLR and NPLR do not have any long-term relationships. Moreover, there is a slight but favorable correlation between NPLR and TLR. This suggests that over the long term, there are no meaningful relationships between the two variables. Nonetheless, there is a substantial positive correlation between NPLR and BS. This demonstrates that over time, a change in NPLR by a certain amount reduces BS. The result demonstrates how the country's banks' sizes are affected by the pace at which the NPLR rises. So, it is crucial to have policies that can aid in recouping all of the loans given to consumers. Last but not least, there is a long-term positive significant relationship between NPLR and LR, showing that a change in NPLR by percentage significantly raises LE.

There was a one-way causal relationship between NPLR and ROA, which is consistent with (Su et al., 2021). The result suggests that NPLR affects ROA. This outcome is not unexpected given that non-performing loans have an impact on a bank's overall assets. The table once again demonstrates the two-way causal relationship between ROA and PLLR that is related to (Shahbaz, 2009; Shahbaz & Islam, 2011). Finally, a unidirectional link between PLLR and NPLR was established.

## Conclusion

This research examines credit risk's impact on the banks' performance. Provision for loan loss was adversely statistically significant, affecting bank performance. This detrimental impact is unavoidable. Loan loss should improve performance. PLLR protects banks against consumer loan losses. Total loan ratio hurt return on asset again. Due to high interest rates, banks may lend without worry. Unpaid debts may be canceled using the interest margin gain.

Yet, credit risk measured by non-performing loan ratio negatively impacted bank performance. This negative impact suggests banks shift losses to other banking services or excessive loan interest rates. NPLs reduce bank performance. Customers will avoid loans with high interest rates.

Due to excessive interest rates, customers will have trouble paying back loans. NPLs will rise again.

## Recommendation

Sierra Leonean banks must follow guidelines while lending to consumers, especially credit officers. The financial industry must provide loans efficiently. They must use a "know your customer" credit procedure (KYC). Use effective credit monitoring and risk controls. Corporate governance reduces credit risk. Again, to improve credit giving, financial institutions should promote financial statement correctness, timeliness, and efficacy.

Further, management must establish efficient and competent risk units to ensure policies are clearly established and conform to authority and limitations. This will lower client loan interest rates. Commercial banks should be driven to recover client loans to decrease bad debt reserves.

Furthermore, management must adopt complicated credit risk mitigation scheme. This includes protecting credit risk, forecasting dubious credits, renegotiating loan conditions for bankrupt clients, third-party credit transfer, extending credit growth, and preplanning customer credit.

Last but not least, comprehensive credit risk policies improve credit risk management rules. Bank management should also lower non-performing loans. The bank will earn if rules are followed and trustworthy clients are found.

Finally, customers seeking loans must follow a credit officer-supervised process. Credit officers who don't cooperate must pay the debt.

## Editors Note

The exploration of the impact of credit risk on bank performance after the World Financial Crisis, evidenced by five commercial banks in Sierra Leone, holds significant implications for the sustainability of the banking sector and the broader economy. Credit risk, exacerbated by economic downturns and financial instability, poses a significant threat to the stability and resilience of banks. By examining how credit risk affects the performance of commercial banks in Sierra Leone post-crisis, this research provide valuable insights into the effectiveness of risk management practices and regulatory frameworks in mitigating financial vulnerabilities and promoting sustainability within the banking sector. Effective risk management strategies enhance stability and resilience of banks, ensuring their ability to withstand external shocks and contribute to economic growth and development. Moreover, by identifying the factors influencing credit risk and its impact on bank performance, policymakers and regulators can formulate targeted interventions aimed at strengthening the financial system, enhancing credit quality, and fostering a conducive environment for sustainable economic development in Sierra Leone. Therefore, this empirical investigation has the potential to inform policy decisions and strategic initiatives aimed at promoting the long-term sustainability of the banking sector and supporting overall economic resilience in Sierra Leone.

## References

Abosedra, S., Shahbaz, M., & Nawaz, K. (2016). Modeling causality between financial deepening and poverty reduction in Egypt. *Social Indicators Research, 126*(3), 955–969.

Afriyie, H. O., & Akotey, J. O. (2013). Credit risk management and profitability of selected rural banks in Ghana. *Journal of Risk Finance, 4*(2), 1–18.

Akhtar, F. M., Ali, K., & Sadaqat, S. (2011). Factors influencing the profitability of conventional banks of Pakistan. *International Research Journal of Finance and Economics, 66*, 117–112.

Alalade, S. A., Binuyo, O. B., & Oguntodu, A. J. (2014). Managing credit risk to optimize banks profitability: A survey of selected banks in Lagos State, Nigeria. *Research Journal of Finance and Accounting, 5*(18), 76–84.

Amidu, M., & Hinson, R. (2006). Credit risk, capital structure and lending decisions of banks in Ghana. *Banks and Banks Systems, 1*(1), 93–101.

Archison, A. (2018). *Credit assessment process and loan repayment: A case study of fidelity bank*. A dissertation submitted to the Department of Finance in Partial Fulfilment of the Requirement for Award of Degree.

Aremu, O. S., Suberu, O. J., & Oke, J. A. (2010). Effective credit processing and administration as a Panacea for non-performing assets in the Nigerian banking system. *Journal of Economics, 1*(1), 53–56. Machiraju, H. R. (2004). *Modern commercial banking*. Vikas Publication, House PVT.

Asare, S. K., Wright, A., & Zimbelman, M. F. (2015). Challenges facing auditors in detecting financial statement fraud: Insights from fraud investigations. *Journal of Forensic and Investigative Accounting*, *7*(2), 63–111.

Athanasoglou, P., Brissimis, S. N., & Delis, M. D. (2005). *Bank-specific, industry specific and macroeconomic determinants of bank profitability*. MPRA Paper No. 153.

Basel Committee. (1999). *Principles for the management of credit risk*. Basel Committee on Banking Supervision.

Basel Committee on Banking Supervision. (2001). *Risk management practices and regulatory capital: Cross-sectional comparison*. Basel: Basel Committee on Banking Supervision, <www.bis.org>

Bashir, A., Khan, M., Sajid, M., & Sorooshian, S. (2023). Basel accord capital regulations and financial risk management: Empirical evidence from Pakistan's financial institutions. *Accounting*, *9*(1), 1–8.

BCBS (1999). A New Capital Adequacy Framework. *Consultative Paper*. <https://www.bis.org/publ/bcbs50.pdf>

Boahene, S. H., Dasa, J., & Agyei, S. K. (2012). Credit risk and profitability of selected banks in Ghana. *Research Journal of Finance and Accounting*, *3*(7), 6–14.

Bourke, P. (1989). Concentration and other determinants of bank profitability in Europe, *North America and Australia. Journal of Banking and Finance 13*, 65–79.

Brewer, E. (1989). Relationship between bank holding company risk and nonbank activity. *Journal of Economics and Business*, 41(4), 337–353. <https://doi.org/10.1016/0148-6195(89)90030-1>

Brewer, E., III, & Mondschean, H. T. (1994). An empirical test of the incentive effects of deposit insurance. *Journal of Money, Credit, and Banking, 26*(1), 146–164.

Brixton, A., Brooks, J., Hecht, P., Ilmanen, A., Maloney, T., & McQuinn, N. (2023). A changing stock—bond correlation: Drivers and implications. *The Journal of Portfolio Management*, Multi-Asset Special Issue 2023, 1–17

Brownbrigde, M. (1998). The causes of financial distress in local banks in Africa and implications for prudential policy. UNCTAD OSG/DP/132. <bsl.gov.sl>.

Cheng, L., Nsiah, T. K., Ofori, C., & Ayisi, A. L. (2020). Credit risk, operational risk, liquidity risk on profitability: A study on South Africa commercial banks. A PLS-SEM analysis. *Revista Argentina de Clínica Psicológica*, *29*(5), 5.

Davies, N. (2023). "Sweet Mother": The neoliberal plantation in Sierra Leone. In *Global plantations in the modern world: Sovereignties, ecologies, afterlives* (pp. 255–283). Cham: Springer International Publishing.

Devinaga, R. (2010). Theoretical framework of profitability as applied to commercial banks in Malaysia. *European Journal of Economics, Finance and Administrative Sciences*. Available at: <http://www.eurojournals.com>.

Do, H., Ngo, T., & Phung, Q. (2020). The effect of non-performing loans on profitability of commercial banks: Case of Vietnam. *Accounting*, *6*(3), 373–386.

Enad, O. M. A., & Gerinda, S. M. A. (2022). Enhancing financial performance of the banks: The role of customer response and operations management. *Journal of Innovation and Entrepreneurship*, *11*(1), 1–16.

Fajembola, O. D., Rahman, N. A. A., & Rus, R. M. (2018). The relationship between the risk management committee and the chief risk officer with the stability of Nigerian banks. *System*, *13*(1), 73–86.

Felix, A. T., & Claudine, T. N. (2008). *Bank performance and credit risk management*. Unpublished Master Dissertation in Finance, University of Skovde, retrieved from <http://www.essays.se/essay/55d54c0bd4/>

Gastineau, G. L. (1992). Option position and exercise limits: Time for a radical change. *Journal of Portfolio Management*, *19*(1), 92.

Gizaw, M., Kebede, M., & Selvaraj, S. (2013). The impact of credit risk on profitability performance of commercial banks in Ethiopia. *African Journal of Business Management*, *9*(2), 59–69.

Godbillon-Camus, B., & Godlewski, C. J. (2005). *Godlewski, Gestion du Risque de Credit dans la Banque: Information Hard, Information Soft et Manipulation*. University Robert Schuman, Strasbourg III.

Greuning, H., & Bratanovic, S. B. (2003). *Analyzing and managing banking risk: A framework for assessing corporate governance and financial risk* (2nd edn). Washington, DC: The World Bank.

Gujarati, D. N., & Sangeetha, S. (2007). *Basic econometric*. New Delhi: McGraw-Hill Education Books Ltd.

Ha, D., & Nguyen, Y. (2023). Institutional quality's influence on financial inclusion' impact on bank stability. *Cogent Economics & Finance*, *11*(1), 2190212.

Haberly, D., & Wójcik, D. (2022). *Sticky power: Global financial networks in the world economy* (p. 400). New York: Oxford University Press.

Haque, A., Mughal, A., & Zahid, Z. (2016). Earning management and the role of accounting conservatism at firm level. *International Journal of Economics and Finance*, *8*(2), 197–205.

Hassid, O. (2008). *La gestion des risques*. Paris: Dunod.

Hiller, J. S., & Jones, L. S. (2022). Who's keeping score? Oversight of changing consumer credit infrastructure. *American Business Law Journal, 59*(1), 61–121.

Hosna, A., Manzura, B., & Juanjuan, S. (2009). *Credit risk management and profitability in commercial banks in Sweden*. Göteborg, Sweden: University of Gothenburg.

Houcine, A., & Houcine, W. (2020). Does earnings quality affect the cost of debt in a banking system? Evidence from French listed companies. *Journal of General Management, 45*(4), 183–191.

Husni, A. K. (2011). Determinants of commercial banks performance evidence from Jordan. *International Research Journal of Finance and Economics,* 149–158. Available at:<http://web.ebscohost.com.www.baser.dk/ehost/pdfviewer/pdf viewer? sid=8b965611- 27b048df- bc38348c87c58664% 40sessionmgr 14&vid =5&hid =14 date.Journal of Finance53, 21072139>

Isenberg, D. T., Sazu, M. H., & Jahan, S. A. (2022). How banks can leverage credit risk evaluation to improve financial performance. *CECCAR Business Review, 3*(9), 62–72.

Jiang, S., Wang, J., Dong, R., Li, Y., & Xia, M. (2023). Systemic risk with multichannel risk contagion in the interbank market. *Sustainability, 15*(3), 2727.

Kaaya, I., & Pastory, D. (2013). Credit risk and commercial banks performance in Tanzania: A panel data analysis. *Research Journal of Finance and Accounting,* ISSN 2222-1697.

Kapur, D., & Gualu, K. A. (2011). What drives the performance of commercial banks in Ethiopia?. *International Journal of Research in Commerce & Management, 2*(7), 1–11.

Kargi, H. S. (2011). *Credit risk and the performance of Nigerian banks,* Department of Accounting, Faculty of Administration, Ahmadu Bello University, Zaria, Nigeria.

Karolina, K. (2023). The influence of return on assets and earnings per share on share price at. Bank Rakyat Indonesia, Tbk. *International Journal of Social and Management Studies, 4*(1), 106–111.

Khalid, S., & Amjad, S. (2012). Risk management practices in Islamic banks of Pakistan. *The Journal of Risk Finance, 13*(2), 148–159.

Kim, D., & Santomero, A. M. (1988). Risk in banking and capital regulation. *The Journal of Finance, 43*(5), 1219–1233.

Koehn, M., & Santomero, A. M. (1980). Regulation of bank capital and ortfolio risk. *Journal of Finance, 35*(5).

Kolapo, T. F., Ayeni, R. K., & Oke, M. O. (2012). Credit risk and commercial banks performance in Nigeria. *Australia Journal of Business and Management Research, 2*(02), 31–38.

Krakah, A.K., Ameyaw, A., & Sällberg, H. (2010). *The Determinants of Bank's Profitability in Ghana: The Case of Merchant Bank Ghana Limited (MBG) and Ghana Commercial Bank (GCB).*

Lee, A., Mitchell, B., & Lederer, A. (2019). *Disinvestment, discouragement and inequity in small business lending.* Washington, D.C.: National Community Reinvestment Coalition. <http://arks.princeton.edu/ark:/88435/dsp01p8418r27s>

Leo, M., Sharma, S., & Maddulety, K. (2019). Machine learning in banking risk management: A literature review. *Risks*, *7*(1), 29.

Mbeba, R. B. (2007). *MFI internal audit and controls trainer's manual.* Microsave-Market-led Solutions for financial services and MEDA. MicroSave Consulting (MSC). <https://www.microsave.org>

Messai, S. A., & Jouini, F. (2013). Micro and macro determinants of non-performing loans. *International Journal of Economics and Financial Issues, 3*(4), 852–860. <https://doi.org/10.1300/J079v27n04_02>.

Metawa, N., Itani, R., Metawa, S., & Elgayar, A. (2023). The impact of digitalization on credit risk: The mediating role of financial inclusion (National Bank of Egypt (NBE) case study). *Economic Research-Ekonomska Istraživanja*, *36*(2), 2178018.

Mohd Amin, S. I., & Abdul-Rahman, A. (2020). The role of regulation in banking: Liquidity risk perspective. *Iranian Journal of Management Studies*, *13*(3), 391–412.

Muhammed, J. K., & Garba, S. (2014). An evaluation of the effects of CRM on the profitability of Nigerian banks. *Journal of Modern Accounting and Auditing, 10*(1), 104–115.

Ogboi, Ch., & Unuafe, O. K. (2013). Impact of credit risk management and capital adequacy on the financial performance of Commercial Banks in Nigeria. *Journal of Emerging Issues in Economics, Finance and Banking, 2*(3), 703–717. Available: <http://www.globalbizresearch.com>.

Ogilo, F. (2012). The impact of credit risk management on financial performance of commercial banks in Kenya. *DBA Africa Management Review, 3*(1), 22–37. <https://doi.org/10.1016/j.enggeo.2011.05.015>.

Opoku, A. (2015). *The impact of credit risk on profitability of some selected Ghanaian banks.* Kwame Nkrumah University of Science and Technology. School of Business, KNUST, College of Humanities and Social Science.

Osei-Assibey, E., & Asenso, J. K. (2015). Regulatory capital and its effect on credit growth, non-performing loans and bank efficiency: Evidence from Ghana. *Journal of Financial Economic Policy*, *7*(4), 401–420. <https://doi.org/10.1108/JFEP-03-2015-0018>

Ozili, P. K. (2017). *Non-performing loans and financial development: New evidence, SSRN.* <https://doi.org/10.2139/ssrn.2892911>.

Ozili, P. K. (2020). 100 quotes from the global financial crisis: Lessons for the future. In *Uncertainty and challenges in contemporary economic behaviour* (pp. 185–194). Leeds: Emerald Publishing Limited.

Pasiouras, F., & Kosmidou, K. (2007). Factors influencing the profitability of domestic and foreign commercial banks in the European Union. *Research in International Business and Finance,* 222–237. Available at: <http://www.sciencedirect.com/science/article/pii/S0275531906000304>. (July 13, 2012).

Pedroni, P. (1995). Panel cointegration. In *Asymptotic and finite sample properties of pooled time series tests with an application to the PPP hypothesis,* Working Papers in Economics, No. 95-013. Indiana: Indiana University.

Pedroni, P. (1999). Critical values for cointegration tests in heterogeneous panels with multiple regressors. *Oxford Bulletin of Economics & Statistics, 61,* 653–670.

Perroti, R. (1993). Political equilibrium, income distribution, and growth. *Review of Economic Studies, 60,* 755–776.

Pham, M. H., & Nguyen, N. M. (2023). Bank funding diversity, risk and profitability: Evidence from Vietnam in the context of the Covid-19 pandemic. *Cogent Business & Management, 10*(1), 2191305.

Phillips, P. C. B., & Perron, P. (1998). Testing for a unit root in time series regression. *Biometrika, 32,* 301–318.

Qi, B., Marie, M., Abdelwahed, A. S., Khatatbeh, I. N., Omran, M., & Fayad, A. A. (2023). Bank risk literature (1978–2022): A bibliometric analysis and research front mapping. *Sustainability, 15*(5), 4508.

Rahman, M. M., Tabash, M. I., Salamzadeh, A., Abduli, S., & Rahaman, M. S. (2022). Sampling techniques (probability) for quantitative social science researchers: A conceptual guideline with examples. *Seeu Review, 17*(1), 42–51.

Richard, E. O., & Okoye, V. I. (2013). Impact of corporate social responsibility on the deposit money banks in Nigeria. *Global Business and Economics Research Journal, 2,* 71–84.

Saeed, M. S., & Zahid, N. (2016). The impact of credit risk on profitability of the commercial banks. *Journal of Business and Financial Affairs, 5*(2), 7–1.

Schaffner, P. (2023). Tiered reserves and trading behavior in the Euro repo market. *Essays in Market Microstructure, 69,* 69–113.

Shahbaz, M. (January, 2009). A reassessment of finance-growth nexus for Pakistan: Under the investigation of FMOLS and DOLS techniques. *The IUP Journal of Applied Economics, 8*(1), 65–80.

Shahbaz, M., & Islam, F. (2011). Financial development and income inequality in Pakistan: An application of ARDL approach. *Journal of Economic Development, 36*(1), 35–58. Seoul: Chung-Ang University Economic Research Institute.

Su, C. W., Qin, M., Rizvi, S. K. A., & Umar, M. (2021). Bank competition in China: A blessing or a curse for financial system?. *Economic Research-Ekonomska Istraživanja, 34*(1), 1244–1264.

Sufian, F., & Chong, R. R. (2008). Determinants of bank profitability in a developing economy: Empirical evidence from the Philippines. *Asian Academy of Management Journal of Accounting and Finance (AAMJAF), 4*(2), 91–112. Available at <http://web.usm.my/journal/aamjaf/vol%2042-2008/42-5.pdf>.

Takyi, L. N., & Naidoo, V. (2022). Is recapitalisation regulation a disrupter? Survival and sustainability of indigenous Ghanaian banks. *Qualitative Research in Financial Markets, 14*(4), 550–569.

Talreja, S., & Hayat, M. (2023). Financial innovation, technological improvement and banks' profitability: An empirical insight. *International Journal of Business and Management Sciences, 4*(1), 87–111.

Triki, O., & Abid, F. (2023). Firms' financial structure with contingent convertible debt, risky debt and multiple growth options. *China Finance Review International, 13*(2), 230–262.

Tung, F. (2020). Financing failure: Bankruptcy lending, credit market conditions, and the financial crisis. *Yale J. on Reg., 37*, 651.

Uhomoibhi, T. (2008). Determinants of bank profitability: Macroeconomic evidence from Nigeria. *Social Science Research Network*, 1–33; Available at: <http://papers.ssrn.com/sol3/papers.cfm?abstract_id=1231064>.

Vermeulen, R., Schets, E., Lohuis, M., Kölbl, B., Jansen, D. J., & Heeringa, W. (2021). The heat is on: A framework for measuring financial stress under disruptive energy transition scenarios. *Ecological Economics, 190*, 107205.

Vong, L. K. (2005). Loans and profitability of banks in Macao. *AMCM Quarterly Bulletin, 2*, 91–107.

Vong, P. I., & Hoi, S. C. (2009). Determinants of bank profitability in Macao. *Macao Monetary Research Bulletin*. Available at <http://www.amcm.gov.mo / publication / quarterly/July2009/macaoprof_en.pdf>.

Winful, E. C., Sarpong, D. J., & Irenee, D. (2020). Financial sector readiness to support economic actives under COVID-19: The case of African continent. *Journal of Economics and International Finance, 12*(4), 164–173. <https://doi.org/10.5897/JEIF2020.1077>.

Xu, M., Wu, W., Li, J., Au, F. T., Wang, S., Hao, H., & Yang, N. (2022). Structural damage detection using low-rank matrix approximation and cointegration analysis. *Engineering Structures, 267*, 114677.

RICHARD SAGOR MITRA AND DR. HEMANT BHANAWAT

# 7 Analyzes of Common Behavioral Bias among Indo-Arabian Investors and How to Better Manage Them

ABSTRACT

Behavioral Finance tries to understand and explain the reasons for irregularities in an investor's investment decision. This study aims to understand the behavioral biases common to Indo-Arabian investors and how finance managers and practitioners can better address these issues to attract potential investors and retain existing ones. The study conducts a systematic literature review to identify common biases and suggest ways to deal with them. The study finds that overconfidence, herding, and loss aversion are common biases displayed by investors in both countries, among other biases. It is suggested that overconfident investors should be provided with correct market information and made aware of the risks associated with unchecked investments. To check herding among investors, companies can customize portfolios in such a way that is suitable for individual customers according to their financial capacities so they become confident to make independent and bold decisions. And, loss averts investors need to be encouraged to see the brighter side of investment so that they are able to balance their emotions and fear of loss. It is recommended to further study the reasons for such similarities in the biases among the investors from a cultural perspective.

Keywords: Behavioral finance, behavioral bias, Indian stock market, Saudi stock market

## Introduction

One of many economic assumptions is the idea of economic man (homo economicus) (Kenton, 2022). The assumption, as described by John Stuart Mill, states that man (humans) is a rational being, whose primary objective is to maximize his satisfaction from the given options (Kapoor

& Prosad, 2017). There are three underlying assumptions to this; perfect information, absolute rationality, and maximizing self-interest. These assumptions lay the foundation for the traditional financial framework. It is sought that while making a financial decision the person has all and perfect information about the market or the options; only makes logical decisions; and, the primary objective of all his decisions is to maximize his satisfaction (maximum utility).

These assumptions help economists and financial practitioners create a theoretical world where the majority of the occupants display the assumed characteristics. This helps them to understand and predict a logical outcome, for a given hypothetical situation. However, the real world does not function that way. In reality, man is not always rational in their financial decisions. Not all of the decisions they make can be logically or mathematically explained. (Although this leads to a rather philosophical question of what is to be considered rational in finance, that is not the scope of this paper.)

Behavioral Finance tries to understand and explain the reasons for irrationalities in the investment decision of investors (Kapoor & Prosad, 2017). One of the classical financial theories is the expected utility theory. This theory holds that when making a decision under uncertainty a person or entity will first weigh (analyze) all the possible outcomes and then choose the course of action that gives the highest utility at any given point in time (Chen, 2022). While the explanation for the expression 'highest utility' from a financial viewpoint means that it brings maximum monetary benefit, in reality, there are other major aspects related to 'maximum utility'.

Behavioral experts refer to these irrationalities as biases. Numerous biases have been identified so far (Kumar & Goyal, 2015; Zahera & Bansal, 2018), with ongoing research in the field. The objective of this research is to identify the common behavioral biases present among investors in India and the Kingdom of Saudi Arabia, respectively. Given that companies depend on investors for raising capital for their operations, it is important that executives and financial practitioners understand the behavioral bias that influences the investors' decision-making process. Hence, the novelty of this study.

The second part of the paper presents a brief review of important concepts in behavioral finance and bias, respectively. The third part presents the Indian and Saudi stock markets and the major biases in each country, respectively. Finally, the paper discusses the common biases found in these two countries and makes suggestions on how to deal with them. The paper ends with indications for further research and a conclusion.

## Behavioral Finance and Biases

Olsen (1998) defined behavioral finance as a "part of science, in that it starts from fundamental axioms and asks whether a theory built on the axiom can explain behavior in the financial marketplace." He also said that behavioral finance is not merely about asserting the rationality of a decision or label them as biases. On the contrary, it should try to "understand and predict systematic financial market implications of psychological decision process."

In their review paper, Ricciardi and Simon (2000) mentioned that ideas of behavioral finance can be traced back to a few original books written back in the 1800s and 1900s. Some of those are MacKay's Delusions And The Madness Of Crowds, Le Bon's The Crowd: A Study Of The Popular Minds, and Selden's Psychology Of The Stock Market, to mention a few. The paper states that psychology, sociology, and finance form the foundations of behavioral finance. This presents the more wholesome inter-disciplinary approach to understanding the decision-making process.

Zahera and Bansal (2018) conducted an extensive revive of literature on the behavioral bias and listed a total of 17 biases. Investors are influenced by various biases while making investment decisions. Those are overconfidence, disposition effect, herding effect, mental accounting, confirmation bias, hindsight bias, house money effect, endowment effect, loss aversion, framing, home bias, self-attribution bias, conservatism, regret aversion, recency, anchoring, and representativeness. These are some of the major biases identified so far with much research being carried out in the field.

Nagy and Obenberger (1994) conducted an empirical study to examine the influence of various utility-maximization and behavioral variables in the decision-making process. They asked the respondents to rank the importance of 34 given variables on the degree of their influence. The findings suggest the classical wealth-maximization criteria has major influences in investor's decision-making process. Top three variables that significantly influence decision-making process are expected corporate earnings, diversification needs, and feelings for the firm's products and service, among others. On the other hand, top three variables that has least influence on investment decision-making are local operations, international operations, and family member opinions, among others.

Since a fuller review of the concept of behavioral finance and its associated biases are not the aim of this paper, we do not present an in-depth review of literature but a short one.

## Research Framework

Employing the systematic literature review method, this research compares and explores the common behavioral biases among investors in India and the Kingdom of Saudi Arabia. Data has been collected from secondary sources only. The first part of the paper will present the behavioral biases displayed by the investors of these countries and identify the common biases. The later part will discuss possible ways to work around those biases.

Comparison of stock markets helps develop a deeper and wider understanding of the investors and the opportunities present in a country. If the companies are aware of the investor's decision-making process, they can customize their offerings to better attract and sustain investors,

### *Choice of Nations*

The bilateral relation between India and Saudi began in 1947. Since then, leaders of both the counties have made continuous effort to strengthen and benefit from the said relation. Back in February 2019, during India visit of Saudi's Crown Prince Mohammad bin Salman, he announced that Saudi has plans of investing approximately $100 billion in India in various sectors, and also signed six MOUs. That same year in October, Prime Minister Narendra Modi visited Saudi and another twelve MOUs were signed between the two nations.[1]

Prominent investment options for Indians in Saudi are Saudi Aramco, SABIC, Alfanar Energy, Public Investment Fund, and Zamil Air Conditioners India Pvt. Ltd, to mention a few. Similarly, prominent investment options for the Arabians in India are Wipro, Tata Group, and Larsen & Toubro, to mention a few.[2]

With such past, ongoing, and future international business collaboration in focus, this study justifies its relevance and importance to understand the behavioral biases that influence their decision making. Because, ultimately it is the investors who will invest in these endeavors and companies need to attract and retain investors.

## India, Indian Stock Market, and Investor's Behavioral Bias

India is one of the largest countries of South Asia and seventh largest in the world. The country gained it's independence in 1947 and is governed by a *Sovereign Socialist Secular Democratic Republic with a Parliamentary system of Government*, that is based on *The Constitution of India*. A total of 28 States and 8 Union Territories makes up the total administrative divisions of the country. The President is the constitutional head of the country, while the Prime Minister is the head of the government.[3]

Historically, archeological art affects confirm that much of the northwestern regions were culturally influenced by the great Indus Vally Civilization. Cultural practices that developed based around this civilization are associated with Hinduism, making it the primary religion in the country. In addition, Buddhism, Jainism, Islam, Christianity is among

1 <https://www.eoiriyadh.gov.in/page/india-saudi-bilateral-relations/>
2 <https://www.investindia.gov.in/country/saudi-arabia>
3 <https://www.india.gov.in/india-glance/profile>

major religions in the country, with many other regional and ethnic religious practices (Schwartzberg, 2023).

One of the most astonishing things about India is its cultural diversity. To begin with, the 28 States have their primary culture as each of these States are made up of their indigenous group of people. For example, the state of Chennai is primarily comprised of Tamilians, Punjab with Punjabis, and West Bengal with Bengalis, among others. Each indigenous group have their very own, distinguished and particular cultural practices and festivals. In general, the daily life-style of people including their financial decisions are highly influenced by their cultural and religious affiliations (Chan & Ananthram, 2019; Schwartzberg, 2023).

There are two major stock exchanges in India, namely, the Bombay Stock Exchange (BSE) and the National Stock Exchange (NSE), both situated in Mumbai. The currency of trading is Indian Rupee. The Securities and Exchange Board of India (SEBI) is the regulatory authority over the stock markets. At both exchanges, trading is facilitated by an open electronic limit order book. India is open to both domestic and foreign investors, subject to various regulations (Singh, 2022).

The BSE was established in 1875 and the NSE was incorporated in 1992, respectively. At present the BSE offers trading in equity, debt instruments, currencies, mutual funds, and derivatives.[4] On the other hand, NSE offers three categories of product, namely, derivatives, fixed income securities and equities.[5] Thousands of well reputed companies listed with both the stock exchanges, respectively.

Studies done on Indian investors reveal the presence of various behavioral biases among investors. Major biases are overconfidence and self-attribution, anchoring bias, mental accounting, the disposition effect, representativeness, herding, emotional biases, loss aversion bias, availability bias, and regret aversion bias, among others (Baker et al., 2019; Chauhan et al., 2020; Prosad et al., 2015; Prosad et al., 2018; Kansal & Singh, 2018; Jain et al., 2020; Mushinada & Veluri, 2018; Kumar et al., 2016).

4 <https://www.bseindia.com/static/about/Company_Overview.html>
5 <https://www.nseindia.com/products-services>

It is important to note that the respondents in these researches are from all over the country. That means, people from various cultural and religious background have displayed these behavioral biases. However, most of the investors are followers of some form of Hinduism or other local religious practice. Demographics presented by the above-mentioned studies state that the number of male traders is more than female traders. The average age group of investors lies between 25 and 45 years old, and mostly college graduates.

## Saudi Arabia, Saudi Arabian Stock Market, and Investor's Behavioral Bias

The modern Kingdom of Saudi Arabia was established by King Abdulaziz Al-Saud in 1935. It is the largest country in the Arabian Peninsula. The country is an Arab and Islamic Sovereign State, with Islam as its state religion and is governed by the Holy Qur'an and the Sunnah as its constitution. The King is the head of the state, who is assisted by the Council of Ministers in discharging his duties.[6]

Historically, the people in these regions lived in tribes, comprising mainly family members. The region has experienced its fair share of wars among these tribes for territory and control of resources. Saudi is the birth place of Islam, hence the primary religion of the country. The cultural and daily life of the people is based on Islamic principles, including food, cloths, and social associations. Beside these, Christians and Hindus form minor religious groups (Teitelbaum et al., 2023).

The Saudi Tadawul Group was established in 2021 and operates the Saudi Stock Exchange (SSE) as its subsidiary.[7] However, the SSE was formed

6 <https://www.saudiembassy.net/history>

7 <https://www.tadawulgroup.sa/wps/portal/tadawulgroup/aboutus/about-us?locale=en>

in 2007 as a joint stock company[8] and is the only stock exchange in the country. The Capital Market Authority (CMA) is the regulatory authority to regulate and develop the stock market in the country. Investment opportunities are available to both national and international investors.[9] At present the exchange offers equities, derivatives, bonds (sukuk) and exchange traded funds (ETF).[10]

Studies reveal the presence of herding behavior, overconfidence, loss aversion, low opportunism, mimetism, sensitivity to rumors, and risk perception (Arabi & Abdelmageed, 2022; Alnajjar, 2013; Alsabban & Alarfaj, 2020; Rahman et al., 2015; Aljughaiman & Chebbi, 2022; Alquraan et al., 2016; Matoussi & Mostafa, 2016). Limited research has been done on the Saudi Stock Market.

Demographical data represent that, male investors are more than female investors. Majority are followers of Islam. The level of education has a positive impact on the quality of their investment decision, while age and gender have no such major influence.

## Discussion

Based on the accessible literature it is observed that overconfidence, herding, and loss aversion are common behavioral biases exhibited by investors in both countries. Due to the limited access to related literature on the Saudi and Indian Stock Markets, respectively, a fuller analysis has not been feasible.

Figure 7.1 shows the network representation showing the interlink among the literature referred in this article. The network points out the common behavioral biases between Indian and Saudi investors. Herding,

8 <https://en.wikipedia.org/wiki/Tadawul>

9 <https://cma.org.sa/en/AboutCMA/Pages/AboutCMA.aspx>

10 <https://www.saudiexchange.sa/wps/portal/saudiexchange/trading/market-services/>

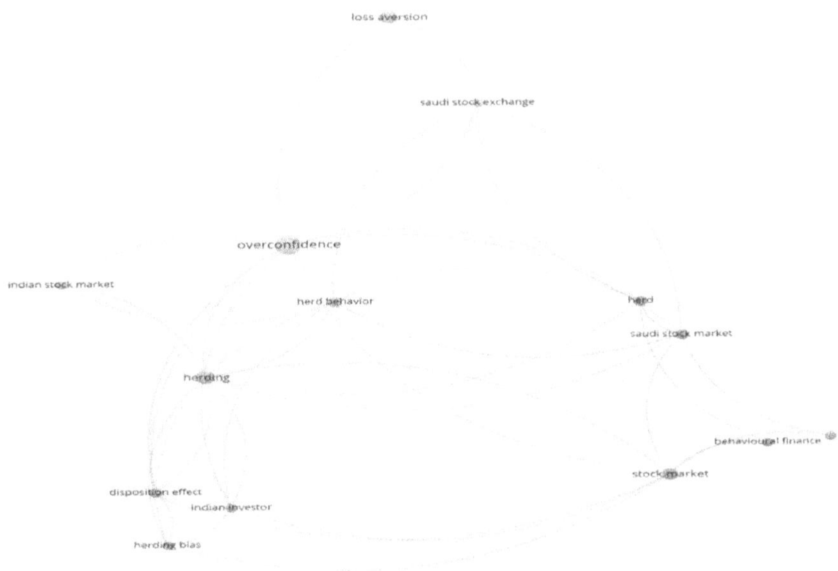


Figure 7.1 Interlink among the literature referred in this article

overconfidence, and loss aversion are the three main biases identified in the literature.

In this section we shall briefly mention the central idea of each of these three biases. In the next section we shall propose practical implications as to how financial practitioners can better manage their public offerings to retain existing and attract potential investors, respectively.

## *Overconfidence*

The term is quite self-explanatory! An investor is said to displaying overconfident bias when they 'overestimate their ability and the accuracy of the information they have' (Byrne et al., 2008). In doing so, they ignore the associated risk (or disregards the level of risk) an investment poses should it fail (Kumar & Goyal, 2015). They attribute the higher performing market to their personal performance and highly optimistic about future market outcomes (Zahera & Bansal, 2018).

Overconfidence can be combined with self-attribution bias, which states that people attribute success to themselves (their skills), but consider it as a bad luck should they fail (Bondt et al., 2008).

Research also suggests that people with less or inadequate information also display overconfidence bias. This is because, almost everyone considers themselves better than an average person and estimate themselves better than how others look upon them. In addition, overconfidence has been described as a *situational phenomenon* that is mainly displayed when people (investors) need to make decisions under uncertainty (Kufepaksi, 2011).

Raheja and Dhiman (2020) suggest that people display overconfidence due to their long years of trading. They believe they are better in their speculations and can become careless about their choices. They assume they will continue to be correct. This means, elderly investors are more likely to display overconfident bias as compared to younger investors.

However, at the same time, the younger investors might be more educated than elderly investors and tech oriented, thereby, consider themselves smarted and display overconfidence. This is a potential idea for future research.

### *Herding*

The simple meaning of the word herding means to move as a group.[11] The term is used to describe similar characteristics in behavioral finance. Investors are said to be displaying a herding bias when a single investor is not able to make their own decision and therefore does what other investors are doing (Raheja & Dhiman, 2020). In many instances, such investors do not make their own investment decisions but take advice and guidance from someone they consider wise or adequately knowledgeable on the matter (Jain et al., 2020).

Herding is also described as a strategy where the new or the middle-class investors closely monitor the investment pattern of the more experienced and wealthy investors, and simply follow their lead. Also, many a

11 <https://dictionary.cambridge.org/dictionary/english/herding>

times investors can simply follow the crowd's investment pattern when they realize the potentiality for general market failure, described as *welfare losses* (Madaan & Singh, 2019). Research also suggests that when people are provided with the information of the decision made by other people, they tend to choose the option with highest selection rate; majority decision (Compen et al., 2022).

### *Loss Aversion*

Loss aversion is one of the core assumptions of the prospect theory that was introduced by Kahneman and Tversky in 1979 (Schmidt & Traub, 2022). The theory states that people choose the option that gives them maximum satisfaction (utility) (Bao & Meng, 2017). When making a decision, "*the absolute subjective value of a specific loss is larger than the absolute subjective value of an equivalent gain*" (Ert & Erev, 2013). Meaning, a loss impacts a person with negative emotions more than the positive emotions if same amount of profit was made. Women are said to be more loss averse than man, while men are more overconfident than woman (Hassan et al., 2014).

## Dealing with Biases

### *Managing Overconfidence Bias*

Investors display overconfidence mainly under uncertainty and lack of proper (authentic) information. Companies need to see that they provide their clients with correct information about their company. Their financial statements should disclose all those figures that are necessary for computing various financial ratios and other market indicators so they can take informed decision instead of assumed decisions. Most Indian investors at least have a college degree (Baker et al., 2019; Sahi & Arora, 2012). Which means they will be able to read and understand various technical aspects of trading if they are provided with the information.

Since overconfident investors tend to ignore the risks associated with their decision, companies can try to explicitly make the investors aware of the risks. In doing so they can aim to gain the investor's trust and a long-lasting relationship. Also, overconfident investors are risk takers and adventurous. Companies can try to create a niche market for that so called "top 1 % rich" into adventurous investment portfolios and harness the positive impact of overconfidence.

It is also found that men display more overconfidence than woman (Jain et al., 2020; Kufepaksi, 2011). Given that both India and Saudi are more of a paternal society, it is no surprise for them to display overconfidence. However, if companies (or brokerage firms) can encourage the investors to develop a practice of discussing their investment plans with their wife, theoretically, a more weighted decision is expected to be made as the overconfident husband will be advised by an underconfident wife who is most likely to point out the risks. Similar idea is proposed by Russo and Shoemaker (1992) who described the process as *counter argumentation*.

Finally, companies ought to focus at increasing the level of financial literacy of the investors. That doesn't mean everyone needs to get a bachelor degree in finance. The simple awareness about the influence of behavioral biases in investment decision making can help investors analyze themselves and avoid displaying such behavior.

### *Managing Herding Bias*

Unlike overconfidence, herding can be a cause of *under-confidence*. Investors might not be confident on the market information or the company's financial statements. Which is why they feel safer with the majority. Providing them with correct and adequate information can provide them with the courage to make their own decisions.

If investors can be provided with combination of investment options that best suits their need and finances, they will not have anyone to follow as those combinations are only made for them. In such scenario they will have to make an independent decision.

Companies can harness the benefits of herding behavior. If a company can attract some major players in the market to invest in their stocks, that could attract other small scale investors. The share price of such a company is most likely to increase. This can be achieved by promoting proper company image, one that is ethical, eco-friendly, humanitarian (SCR activities), among others.

Herding can prove to be dangerous for the company if investors collectively decide to boycott investing in a particular company. There is a high possibility of witnessing this phenomena in Saudi, as there are certain products considered *haram* (sin product) according to the teachings of Islam (Alam et al., 2017). Similarly, many Indians are pro nature and prefer to invest in 'eco-friendly' companies these days, and boycott companies that they consider are harming nature. Companies should consider the socio-religious background of the investors.

### *Managing Loss Aversion Bias*

The loss aversion bias is very subjective and hence it is difficult to suggest any technique to resolve it. People perceive happiness and sadness at different levels. Research shows that loss aversion can occur at both risky and riskless situations. Loss aversion increases with age, income and wealth (Gächter et al., 2022). This implies that young investors are less loss averse. The demographic information on these two countries shows that most of the investors are male and age between 25 and 35. Companies will need to prove themselves to attract investors as younger investors are tech-and-info savvy.

Companies need to project the benefits more than the potential loss (risk). However, that doesn't mean the company should trick the investors by selling lies or false hope. Loss averse investors need to be offered with investments with sure return.

## Conclusion and Recommendations for Further Studies

Behavioral biases are global and unavoidable. Unless humans become cyborgs, there will always be some or the other kinds of bias in their decision-making, because, after all, emotion is what makes us human. This paper has identified the various behavioral biases displayed by Indian and Saudi Arabian investors, respectively, from existing literature. Among many, overconfidence, herding and loss aversion are common biases noticed among investors of both the country. While biases can't be fully eradicated measures can be taken to reduce their impact in investment decision process. Financial literacy lies at the core in this endeavor (Adil et al., 2022). The simple awareness about these biases and how they are displayed rather unconsciously can help an investor analyze themselves to identify these biases and take corrective measures whenever necessary (Baker & Ricciardi, 2014). In addition, companies need to be mindful of the socio-cultural and religious affiliation of their target customers. They will be able to serve their clients better when they are aware of the biases that influence their decision making (Raheja & Dhiman, 2020).

Further study is recommended on the influence of culture on investors behavioral bias.

## Editors Note

The analysis of common behavioral biases among Indo-Arabian investors and strategies for better managing them holds significant importance for sustainability in financial markets and investment practices. Behavioral biases, such as overconfidence, herding behavior, and loss aversion, can lead to suboptimal investment decisions, market inefficiencies, and systemic risks. By understanding the prevalence and impact of these biases among Indo-Arabian investors, this research offer valuable insights into improving investment decision-making processes and promoting financial stability. Implementing effective risk management strategies, investor education programs, and regulatory interventions can help mitigate the adverse effects of behavioral biases, thereby enhancing market transparency, investor confidence, and overall sustainability in the financial sector. Moreover, by fostering a culture of informed decision-making and risk awareness, stakeholders can contribute to the resilience and long-term viability of financial markets, ensuring their ability to support economic growth and development in the Indo-Arabian region and beyond. Therefore, this analysis has the potential to inform policy initiatives and industry practices aimed at promoting sustainable investment behavior and mitigating systemic risks in financial markets.

## References

Adil, M., Singh, Y., & Ansari, M. S. (2022). How financial literacy moderate the association between behaviour biases and investment decision? *Asian Journal of Accounting Research*, *7*(1), 17–30. <https://doi.org/10.1108/AJAR-09202000086>

Alam, Md. M., Akbar, C. S., Shahriar, S. M., & Elahi, M. M. (2017). The Islamic Shariah principles for investment in stock market. *Qualitative Research in Financial Markets*, *9*(2), 132–156. <https://doi.org/10.1108/QRFM-0920160029>

Aljughaiman, A. A., & Chebbi, K. E. (2022). Do investor overconfidence and loss aversion drive Saudi firm market performance? The moderating effect of corporate governance. *Sustainability (Switzerland)*, *14*(16). <https://doi.org/10.3390/su141610072>

Alnajjar, M. I. M. (2013). Behavioral inferences of Tadawul investor. *International Journal of Business and Management*, *8*(24), 17–24. <https://doi.org/10.5539/ijbm.v8n24p17>

Alquraan, T., Alqisie, A., & Al Shorafa, A. (2016). Do behavioral finance factors influence stock investment decisions of individual investors? (Evidences from Saudi Stock Market). *American International Journal of Contemporary Research*, *6*(3), 159–169. <www.aijcrnet.com>

Alsabban, S., & Alarfaj, O. (2020). An empirical analysis of behavioral finance in the Saudi Stock Market: Evidence of overconfidence behavior. *International*

*Journal of Economics and Financial Issues, 10*(1), 73–86. <https://doi.org/10.32479/ijefi.8920>

Arabi, K. A. M., & Abdelmageed, H. M. (2022). Is Herd behavior still persistent in the Saudi Stock Market? *Archives of Business Research, 10*(2), 244–251. <https://doi.org/10.14738/abr.102.11886>

Baker, H. K., Kumar, S., Goyal, N., & Gaur, V. (2019). How financial literacy and demographic variables relate to behavioral biases. *Managerial Finance, 45*(1), 124–146. <https://doi.org/10.1108/MF-0120180003>

Baker, H. K., & Ricciardi, V. (2014). How biases affect investor behaviour. *The European Financial Review*, 7–10. <https://www.researchgate.net/publication/280087380_How_Biases_Affect_Investor_Behaviour>.

Bao, H. X. H., & Meng, C. C. (2017). Loss aversion and housing studies. *Journal of Real Estate Literature, 25*(1), 49–75. <https://doi.org/10.1080/10835547.2017.12090454>

Bondt, W. F. M. De, Muradoglu, Y. G., Shefrin, H., & Staikouras, S. K. (December 3, 2008). Behavioral finance: Quo Vadis?. *Journal of Applied Finance (Formerly Financial Practice and Education), 18*(2). Available at SSRN: <https://ssrn.com/abstract=2698614>

Byrne, A., Brooks, M., & Gifford, B. (2008). Behavioral finance: Theories and evidence. *Behavioral Finance: Theories and Evidence*, 1979, 1–26. <https://doi.org/10.2470/rflr.v3.n1.1>

Chan, C., & Ananthram, S. (2019). Religion-based decision making in Indian multinationals: A multi-faith study of ethical virtues and mindsets. *Journal of Business Ethics, 156*(3), 651–677. <https://doi.org/10.1007/s1055101735587>

Chauhan, Y., Ahmad, N., Aggarwal, V., & Chandra, A. (2020). Herd behaviour and asset pricing in the Indian stock market; Herd behavior and asset pricing. *IIMB Management Review, 32*(2), 143–152. <https://doi.org/10.1016/j.iimb.2019.10.008>

Chen, J. (October 14, 2022). Expected utility: Definition, calculation, and examples. *Investopedia*. Retrieved March 21, 2023, from <https://www.investopedia.com/terms/e/expectedutility.asp>

Compen, B., Pitthan, F., Schelfhout, W., & De Witte, K. (2022). How to elicit and cease herding behaviour? On the effectiveness of a warning message as a debiasing decision support system. *Decision Support Systems, 152*(August), 113652. <https://doi.org/10.1016/j.dss.2021.113652>

Ert, E., & Erev, I. (2013). On the descriptive value of loss aversion in decisions under risk: Six clarifications. *Judgment and Decision Making, 8*(3), 214–235. <https://doi.org/10.1017/s1930297500005945>

Gächter, S., Johnson, E. J., & Herrmann, A. (2022). Individual-level loss aversion in riskless and risky choices. *Theory and Decision*, *92*(3–4), 599–624. <https://doi.org/10.1007/s11238021098398>

Hassan, T. R., Khalid, W., & Habib, A. (2014). Overconfidence and loss aversion in investment decisions: A study of the impact of gender and age in Pakistani perspective. *Research Journal of Finance and Accounting*, *5*(11), 148–158. <www.iiste.org>

Jain, J., Walia, N., & Gupta, S. (2020). Evaluation of behavioral biases affecting investment decision making of individual equity investors by fuzzy analytic hierarchy process. *Review of Behavioral Finance*, *12*(3), 297–314. <https://doi.org/10.1108/RBF-0320190044>

Kansal, P., & Singh, S. (2018). Determinants of overconfidence bias in Indian stock market. *Qualitative Research in Financial Markets*, *10*(4), 381–394. <https://doi.org/10.1108/QRFM-0320170015>

Kapoor, S., & Prosad, J. M. (2017). Behavioural finance: A review. *Procedia Computer Science*, *122*, 50–54. <https://doi.org/10.1016/j.procs.2017.11.340>

Kenton, W. (2022, December 13). What is an economic man? Definition, use in analysis, and history. *Investopedia*. Retrieved March 21, 2023, from <https://www.investopedia.com/terms/e/economic-man.asp>

Kufepaksi, M. (2011). The impact of overconfidence and gender on trading activity: How well do investors assess probability? *World Review of Entrepreneurship, Management and Sustainable Development*, *7*(4), 440–462. <https://doi.org/10.1504/WREMSD.2011.042895>

Kumar, A., Bharti, M., & Bansal, S. (2016). An examination of herding behavior in an emerging economy—A study of Indian stock market. *Global Journal of Management and Business Research, 16*(5), 1–9.

Kumar, S., & Goyal, N. (2015). Behavioural biases in investment decision making—A systematic literature review. *Qualitative Research in Financial Markets*, *7*(1), 88–108. <https://doi.org/10.1108/QRFM-0720140022>

Madaan, G., & Singh, S. (2019). An analysis of behavioral biases in investment decision-making. *International Journal of Financial Research*, *10*(4), 55–67. <https://doi.org/10.5430/ijfr.v10n4p55>

Matoussi, H., & Mostafa, A. (2016). *An investigation of investors bias on Saudi stock exchange*, *Economic Research Forum*, 1–26. <https://erf.org.eg/app/uploads/2018/02/Matoussi-and-Mostafa.pdf>.

Mushinada, V. N. C., & Veluri, V. S. S. (2018). Investors overconfidence behaviour at Bombay stock exchange. *International Journal of Managerial Finance*, *14*(5), 613–632. <https://doi.org/10.1108/IJMF-0520170093>

Nagy, A., & Obenberger, W. (1994). Influencing individual behavior investor. *Financial Analysts Journal*, *50*(4), 63–68.

Olsen, R. A. (1998). Behavioral finance and its implications for stock-price volatility. *Financial Analysts Journal, 54*(2), 10–18. <http://www.jstor.org/stable/4480062>

Prosad, J. M., Kapoor, S., & Sengupta, J. (2015). Behavioral biases of Indian investors: A survey of Delhi-NCR region. *Qualitative Research in Financial Markets, 7*(3). <https://doi.org/10.1108/QRFM-04201400012>

Prosad, J. M., Kapoor, S., Sengupta, J., & Roychoudhary, S. (2018). Overconfidence and disposition effect in Indian equity market: An empirical evidence. *Global Business Review, 19*(5), 1303–1321. <https://doi.org/10.1177/0972150917726660>

Raheja, S., & Dhiman, B. (2020). How do emotional intelligence and behavioral biases of investors determine their investment decisions? *Rajagiri Management Journal, 14*(1), 35–47. <https://doi.org/10.1108/ramj-1220190027>

Rahman, M. A., Chowdhury, S. S. H., & Shibley Sadique, M. (2015). Herding where retail investors dominate trading: The case of Saudi Arabia. *Quarterly Review of Economics and Finance, 57*, 46–60. <https://doi.org/10.1016/j.qref.2015.01.002>

Ricciardi, V., & Simon, H. K. (2000). What is behavioral finance?. *Business, Education & Technology Journal, 2*(2), 1–9. Available at SSRN: <https://ssrn.com/abstract=256754>

Russo, E. J., & Shoemaker, P. J. (1992). Managing overconfidence. *Sloan Management Review, 33*(2), 7–17.

Sahi, S. K., & Arora, P. A. (2012). Individual investor biases: A segmentation analysis. *Qualitative Research in Financial Markets, 4*(1), 6–25. <https://doi.org/10.1108/17554171211213522>

Schmidt, U., & Traub, S. (2022). An experimental test of loss aversion. *Journal of Risk and Uncertainity, 25*(3), 233–249. <https://www.jstor.org/stable/41761121>

Schwartzberg, J. E., Calkins, P. B., Subrahmanyam, S., Thapar, R., Wolpert, S. A., Raikar, S. Pai, Alam, M., Srivastava, A. L., Dikshit, K. R., Spear, T. G. P., Allchin, F. R., & Champakalakshmi, R. (March 25, 2023). *India: Encyclopedia Britannica*. <https://www.britannica.com/place/India>

Singh, M. (July 13, 2022). An introduction to the Indian stock market. *Investopedia*. Retrieved March 31, 2023, from <https://www.investopedia.com/articles/stocks/09/indian-stock-market.asp>

Teitelbaum, J., Ochsenwald, W. L., & Philby, H. St. J. B. (2023, March 25). *Saudi Arabia: Encyclopedia Britannica*. <https://www.britannica.com/place/Saudi-Arabia>

Zahera, S. A., & Bansal, R. (2018). Do investors exhibit behavioral biases in investment decision making? A systematic review. *Qualitative Research in Financial Markets, 10*(2), 210–251. <https://doi.org/10.1108/QRFM-04201700028>

AKASH KUMAR BISWAL AND BISWAJIT PATRA

# 8 Determinants of Pensions: Empirical Evidence from OECD Countries

ABSTRACT

Purpose: To analyze the macro determinants of pension for the OECD countries.

Methodology: This study uses the second-generation panel estimation methods: Fixed effect Model with Driscoll-Karry standards errors and Bootstrapped panel quintile regression over the data from 2010 to 2020 to find the effects of socioeconomic variables on the pension system in the OECD countries.

Findings: The results of the fixed effect model show that the per capita GDP, health expenditure, institutional quality index, and government debt are positively and significantly associated with the pension system in the OECD countries. In contrast, life expectancy does not significantly affect the pension system. Furthermore, the Bootstrapped panel quintile regression results also support the results of the fixed effect model.

Limitations: The study could not capture the information on pension expenditure for private and government employees separately due to the unavailability of data.

Implications of the study: The study suggests that the lower and medium pension expenditure countries spend more on a pension so that, the quality of retired people's life increases.

Originality: To the authors' best knowledge, this paper is the first of its kind to do an empirical study that has investigated the determinants of pension in OECD countries with recent data.

Keywords: Pension system, OECD, GDP per capita, fixed effect, bootstrapped panel quintile regression

JEL Classification: H55; H75

## Introduction

The pension system of a country significantly impacts its social welfare (European Commission, 2012). Help Age International also mentions it is a primary social safety used to shield people from the socioeconomic dangers and vulnerabilities of aging because the pension system is a viable, cost-effective option for achieving economic stability and lowering poverty and inequality in later life (Juergens, 2020). Using international human rights criteria and values, as well as the social pensions system, everyone has the right to adequate social security and protection at all stages of life (International Human Rights Law, 2019). In earlier studies, Floristeanu (2013) demonstrated both social (such as political, demographic, and cultural) and economic variables such as economic growth, budget deficits, debt loads, financial volatility, and low occupancy (Mfoutou & Danquah, 2023) impact the pension system of a nation. According to a report by HelpAge International (Juergens, 2020), health expenditure positively impacts the pension system. Some literature finds similar results in the association between health spending and the public pension system. In this context, An et al. (2016) conducted a study on OECD nations over the data from 1980 to 2005 and demonstrated that health spending significantly impacts the public pension system. The study also found that the increase in public pension programs leads to a rise in total health spending.

Further, it has been established that older individuals have higher health costs than younger ones, which results in significant out-of-pocket health spending in most nations. Thus, the primary objective of pension systems is to offer adequate retirement incomes, access to a high standard of living, and financial independence for older people. Moreover, it is observed from the existing literature, the following are the primary goals of the pension system: to end elder poverty and provide at least a basic level of financial stability to the elderly (Juergens, 2020). Besides, it reduces inequality in old age and redistributes money between the wealthy and the poor and between the younger and older generations. Also, it offers insurance to the people so they can be protected against the possibility of outliving their savings and getting a steady income stream for the remainder of their lives. Moreover, (Kaganovich & Zilcha, 2008) categorized the pension systems into two plans: a pay-as-you-go plan and a defined contribution (DC) or fully funded (FF) plan. In the fully funded (FF) system, the pensioners are the sponsor, whereas the pay-as-you-go scheme distributes pension payments to the old by taxing the young.

This paragraph discusses the overview of the pension system in the OECD countries. The development of pension systems in OECD countries is linked with the industrialization era which led to the emergence of wage labor and the need for social security. The first pension system was introduced in Germany in 1889, followed by other European countries, including France, Denmark, and Austria. The pension system in the United States was recognized in 1799 with the enactment of the Social Security Act (Clark et al., 1999). This program provided retirement benefits to workers aged 65 and older and was financed through payroll taxes. However, some significant changes have been made to pension systems in many OECD countries in recent years. These amendments are made to deal with issues like growing older populations, longer life spans, and the inability of pension systems to keep up with inflation. According to an OECD report (OECD, 2021), one of the major contests confronting pension systems in OECD nations is the demographic shift toward an aging population, which results in an increase in the number of retirees relative to the number of workers contributing to the pension system. This has put pressure on pension funds and raised concerns about the sustainability of pension systems. Another challenge is the increase in life expectancy, which has led to longer periods of retirement and increased demand for pension benefits. In addition, low-interest rates and economic uncertainty have affected the performance of pension funds, making it more difficult for pension systems to meet their financial obligations. In this context, a study conducted by Kuitto et al. (2023) found that old-age pensions and retirement income arrangements play a vital role in OECD countries to maintain a suitable level of life after retirement. They found that pensions help to increase the happy life of old people. Further, it is found that in many nations, public pension systems are struggling, whereas, in some countries, retirees benefit significantly from the pension. The OECD countries have achieved a decline in the prevalence of poverty among older people; however, some retirees still worry about it because of complicated layers of provision, the rules and regulations, and the interaction between public and private pension systems which make retirement income arrangements more complex in the OECD nations. So far the age limit is concerned for availing the public pension benefits is about 65 years and the government bears the full residential cost of the eligible people in the OECD nations (Kalish & Aman, 1998).

This study tries to find out the significant parameters that impact pensions in the OECD economies. In other words, it attempts to analyze the impact of per capita GDP, institutional quality index (IQI),[1] health expenditure, central government debt, and life expectancy on the pension in the panel of 38 OECD countries over the data from 2010 to 2020. It uses panel data techniques for analysis. The paper has gone through several pre and post-diagnostic checks before finalizing the model. Tests such as panel unit roots, Hausman test, BP LM Test for random effect model, autocorrelation, heteroscedasticity, cross-sectional dependence, and cointegration are carried out for the analysis. Based on the prescriptions of the test the paper has selected the Fixed effect Model with the Driscoll-Karry standards errors for the analysis. Further, to check how the impact varies in relation to different groups of pensions giving economies, Bootstrapped panel quintile regression (BPQR) has been carried out. This helps to distribute the economies into different quantiles based on the amount of pension spending as a percentage of GDP. This tries to show the results considering the heterogeneity among the selected economies. Further, it shows how the socioeconomic indicators impact the pension system of different quantiles of OECD countries.

There are only a few studies (such as Floristeanu, 2013; An et al., 2016; Kuitto et al., 2023) available that give theoretical support to the factors that affect the pension system of a country. Presently, there are no well-established empirical findings, to say what influences the pension in the

1 IQI represents the constructed variable from the world governance indicators available from world bank research for the institutional standard of Individual countries. The country having better institutions have better score in the IQI.

OECD economies. This creates a motivation to check the determinants of pension empirically. This paper also incorporates how socioeconomic variables affect that pension. Therefore, this study is novel in this direction.

The study has made a significant contribution to the existing literature in the following ways. First, it generates novel empirical analysis for the OECD economies, considering the determinants of pension. Secondly, it investigates how the above-mentioned variables affect the pension system in OECD countries by following the second-generation panel estimation procedure. Third, advanced panel estimation techniques such as Bootstrapped panel quintile regression and the Fixed effect Model with the Driscoll-Karry standards errors model have been used to address issues like cross-sectional dependence, autocorrelation, and heteroscedasticity and give a robust result. Further, the empirical conclusions of this study will provide a roadmap to the policymakers and stakeholders of the sampled countries. The rest of the study is designed as follows: Section "Review of Literature" deals with the literature review, research gap, theoretical underpinnings, and hypothesis consigning. Section "Framework for Data and Models" discusses the data and methodology used and section "Econometric Method" represents the empirical findings and discussion. Finally, section "Results and Discussion" offers the concluding remarks, policy implications, limitations, and future research avenues.

## Review of Literature

The OECD is comprised of the world's 38 most advanced countries. The pension systems of the OECD countries are affected by a variety of economic and social factors (Floristeanu, 2013). It gets influenced by the fiscal policy and policy-making institutions of the countries. In a similar line, for 25 OECD nations, An et al. (2016) and Azolibe (2022) looked at how health spending affects pension systems favorably. They made use of the panel data models and used the data period between 1980 and 2005 for the analysis. They discovered that for every percentage point increase in pension, health expenses rise by 3 %. It shows there is a direct relationship between health spending and pension spending for the OECD economies. Further, Been et al. (2017) explored how the pension system reduces income disparity and old-age poverty in the European economies. Another study (Kuitto et al., 2023) in a similar fashion on pension for the OECD economies, found that pension helps to reduce social issues like poverty. They found that the public pension system is lowering the rate of old-age poverty in 14 developed OECD nations. They used data from the period 1980 to 2010 and samples of 14 advanced OECD nations to support their result. Following this, Holzner et al. (2022) studied the 35 OECD economies and used panel data models. They found that pension does not get influenced by the volatilities arising from macroeconomic situations of the economies. They noted that volatilities in GDP, consumption, and investment of economies do not impact the pension for the economies.

A nation's social well-being is greatly influenced by its pension system (European Commission, 2012). According to Floristeanu (2013) and Alon (2014), a country's pension system is affected by a variety of socio-economic factors. They discovered that the pension system is hindered by slow economic development, budget deficits, debt loads, unstable financial conditions, and low occupancy. They also discovered that political, demographic, and cultural factors influence pensions. Pension also affects the mental health of people. Mostert et al. (2022) in their paper analyzed how pensions affected depression, traumatic stress, and death of people after the 2008 pension reforms in South Africa. They discovered that the pension system decreased depression, traumatic stress, and depth by 3 %, 4 %, and 5 %, respectively for the people of South Africa. Like this, Hwang and Lee (2022), revealed that the social pension system significantly enhances mental and physical agility. The grip strength increased by 2.3 kg after five years of receiving the social security pension. Additionally, the social pension may have enhanced the recipients' eating practices, use of healthcare, and level of physical activity while lowering their financial burden and related stress. Social pensions reduce the usage of solid fuels by rural residents. Family members' religious membership magnifies the negative effects of the social pension on household solid fuel usage. The negative effects of social pensions on household solid fuel usage are amplified by the percentage of racial or ethnic minorities in a family (Li et al., 2022). The social pension has a positive influence on spending among the less fortunate population because its benefits are more obvious among single people than in families with other people (Kang et al., 2022).

The pension one of the key components of the social welfare system. The German social pension security system benefits legal workers more than self-employed workers, who do not contribute to the country's fiscal sustainability (Heer & Trede, 2023). Public sector employees get better pension benefits compared to employees working in the private sector. In China to analyze, the pension gap between the public and non-public sector, Wang et al. (2023) used the unconditional quantile regression developed by Firpo et al. (2009) and found that the public sector retirees get more pension rewards over their counterparts in the nonpublic sectors. The pension system in Korea affects health and well-being in old age. They formulated that constantly receiving social pensions significantly enhanced receivers' control power and cognitive ability of old age people (Hwang & Lee, 2022). When compared, to the public sector employees to their nonpublic sector peers, retirees in the public sector enjoy better pension benefits (Wang et al., 2023). The ways in which a pension provider could prevent prospective pension reductions during the decumulation period from harming their reputation, to address this Penas et al. (2023) suggested that the asset value should always exceed the net present value of the pension liabilities to identify the optimal course of action in each case to maintain the viability of the pension plan. Another study, by Joubert and Todd (2022) on the gender disparity for pension found that the 2008 pension reform was mainly effective in eradicating gender inequities and reducing old-age poverty in Chile. They found that three design features that significantly diminish gender pension discrepancies include elevating women's retirement age to the same level as men's retirement age, escalating minimum pension advantage eligibility, and providing a per-child pension incentive. One of the recent studies by Romp and Beetsma (2022), studied the pension for the OECD economies and found that the demographic dividend of an economy determines the pension reforms and they also noted that the business cycle determines the timing for the implementation of the pension reform. Brown and Fraikin (2022) studied the household replacement rate for the old age pension. They found that the household replacement rate has a key influence in lowering the senior poverty rate because houses with asymmetrical employment arrangements usually fall in the lowest section of the distribution of equivalized income. The basic pension system (BPS) increased expenditure on basics like food and clothes, as well as on healthcare, but not on leisure due to the growth in the ratio of total spending to absolute and relative poverty levels.

Following these studies, we observe that there is no clear evidence on the determinants of pension. Several articles are there on pension studies, however, the socioeconomic aspect has not been considered in the analysis. This paper tries to fill the gap in this respect. This paper tries to use the recently available data till the end of 2020 for analyzing the major factor which affects the pension for the OECD economies. We have considered a set of socio-economic factors for finding out the determinants of pension. This study helps to identify the important factors which may influence the pension for the OECD economies. For the empirical analysis, this paper uses data captured from 38 OECD nations. The data spans from 2010 to 2020. In this study, we used the five independent variables known as per capita GDP, institutional quality index, health expenditure, central government debt and life expectancy, and pension as dependent variables. In this study, we defined how social and economic factors determine the pension system. In other words, we can say how the independent variable positively or negatively affects the pension system in the OECD countries. To find the answer to this question, we have used the fixed effect Driscoll standards error model and bootstrapped panel quantile regression econometric model.

## *Hypothesis Development*

The pension systems of various nations may be impacted by social and economic variables. Theoretically, this study's findings are consistent with those of Ragner Nurkse (Vicious Circle, 1953), who found that as a nation's GDP per capita increases and its citizens enjoy better health, the nation's life expectancy rises. If a nation's life expectancy is high, its elderly population will increase, and the government will offer more pensions for social welfare. When people receive more pensions, their income level rises as well, increasing GDP per capita (Floristeanu, 2013). Again, related occurrences also took place.

On the basis of existing literature, we have framed the theoretical framework shown in Figure 8.1, which shows a cyclical relationship among GDP per capita, health expenditure, life expectancy, and pension system. It shows that a rise in the GDP per capita of countries causes an increase in health expenditure, it is believed that a rise in health expenditure reduces health-related issues which ultimately increases life expectancy. The life expectancy of a country is high, then the more aged people are living in these countries. To increase the social welfare of the countries, the government should provide more pensions. If the pension amount of countries increases, the GDP per capita will increase as the pension is a source of income for them (Juergens, 2020).

A country's social welfare is significantly impacted by its pension system, which also improves governance and raises the happiness index. The standard of a nation's institutions also has an impact on its pension system. The pension system of a nation is impacted by a variety of factors,

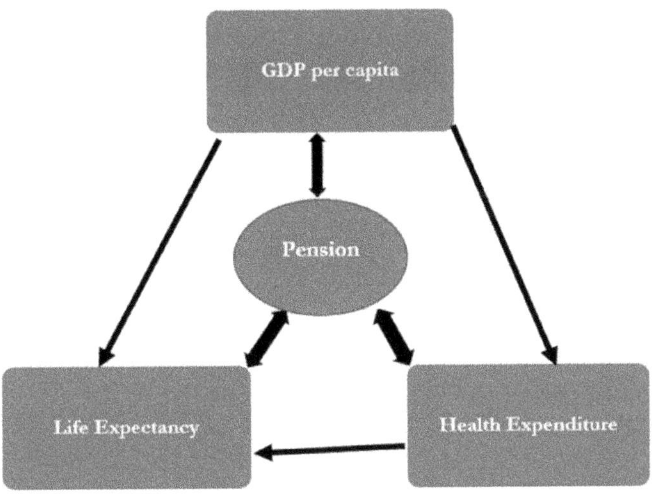


Figure 8.1: Theoretical framework

including the GDP per capita, health care expenditures, national debt, and life expectancy. This study examines the nexus between the effects of Economic growth, institutional quality, healthcare spending, national debt, and life expectancy on pensions expenditure.

As per the available literature, we have found that there is less relevant document available on the relationship between per capita GDP, Institutional quality index, Health Expenditure, Central Government debt, and life expectancy with pension expenditure. Therefore, this study attempts to fulfill this gap by studying the relationship between these variables of the OECD countries. Before the empirical investigation, this study frames the following hypothesis:

**Hypotheses 1:** Association between per capita GDP and Pension

The per capita GDP of a country may have a significant influence on its social welfare system. A country will aim to improve its social welfare if its GDP per capita is high. As a result of governments spending more on social security, pensions naturally rise. As a country's pension grows, so does its GDP per capita, and this cycle repeats (Holzner et al., 2022). Figure 8.2 depicts the trends between pension and GDP per capita. This trend shows a direct relationship between them, but between 2018 and 2019 the pension trends so little decline because at that time some OECD countries (Japan, Germany, and Indonesia) changes their age structure

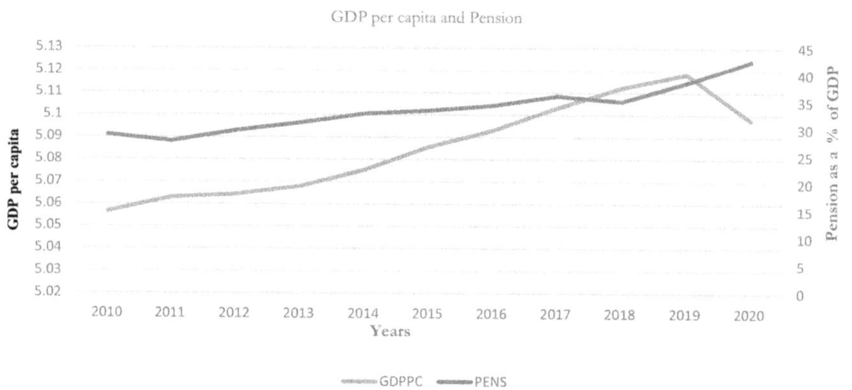


Figure 8.2: Trends between pension and GDP per capita in OECD countries

for pensions. And in the case of GDP per capita, it suddenly decreases after 2019 because of mainly three reasons: (I) Implementation of tariffs by major economies, particularly the United States, and retaliatory actions taken by others like a chain. So, industrial production and trade have declined and business confidence has fallen. (II) In 2018, the US economy grow quickly because of expenditure increases and tax cuts that boosted demand. As a result, the US federal reserve has kept rising the policy interest rate. And the last one is (III) The US dollar has risen versus most of the currencies in 2018 as growth and interest rate in the US have exceeded those in other major economies.

Following this, the present paper suggests checking whether the rise in GDP per capita increases pension spending for the OECD economies. This has led to the following hypothesis.

$H_0$: Higher GDP per capita does not increase the Pension expenditure
$H_1$: Higher GDP per capita does increase the Pension expenditure

**Hypothesis 2:** The association between institutional quality index (IQI) & pension.

The IQI is constructed by using the following sector: (i) voice and accountability; (ii) political stability and absence of violence or terrorism; (iii) government effectiveness; (iv) regulatory quality; (v) rule of law; and (vi) control of corruption. Therefore, if a country's institutional quality index is high, its social welfare system and pension system are expected to be high. The study (Floristeanu, 2013) found that countries with better institutional quality spend more on social security and pension. Following this, the present paper tries to check whether countries having better institutions spend more on the pension system and frames the following hypothesis as mentioned below:

$H_0$: Higher institutional quality index of a country does not increase the Pension expenditure
$H_1$: Higher institutional quality index of a country increases the Pension expenditure.

**Hypothesis 3:** The association between health expenditure and pension.

A nation's health spending will most likely increase as its elderly population grows. Elderly people require higher health spending due to the high rates of illness and impairment experienced by them. This will cause high levels of health expenditures in many nations. An et al. (2016) in their study noticed that higher health spending led to a rise in pension spending for the OECD nations. Therefore, the present hypothesis examines whether health spending positively impacts the pension system.

The above Figure 8.3 shows that there is a direct relationship existed between pension and health expenditure. But between the period from 2019 to 2020, health expenditure increases more proportionally than pension because of the Covid-19 pandemic. This study also examines how it varies for different quantiles of pension spending economies. With this background, the present paper examines the following hypothesis.

$H_0$: Higher health expenditure in a country does not increase Pension spending.
$H_1$: Higher health expenditure in a country increases Pension spending.

**Hypothesis 4:** The association between central government debt, life expectancy, and pension.

The pension system helps people favorably in high-income nations and impacts negatively in low-income countries. The rise in central government debt may put pressure on the government to spend on social security measures like pensions. Higher pension spending may also cause fiscal stress for the economy. Hence, the debt of the government may

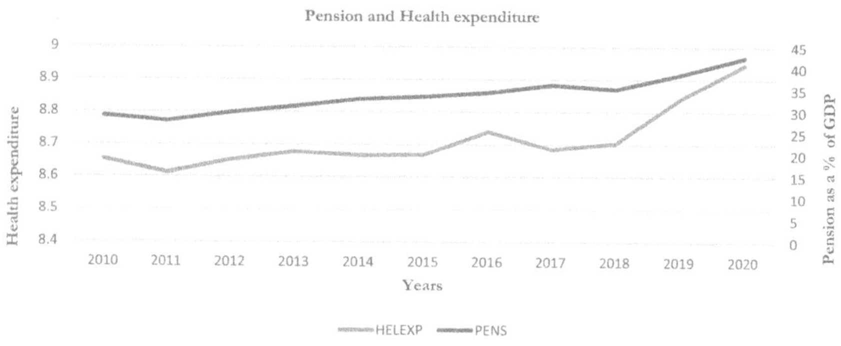


Figure 8.3: Trends between pension and health expenditure in OECD countries

influence pension spending. A rise in the health expectancy of a nation is an indicator to show people are having a longer lifespan. This may increase the pension spending for the people.

Figure 8.4 shows the direct relationship between pension and central government debt of the OECD countries. The pension increases with the central government debt but after 2019 the speed of the increase in central government debt is more than the speed of the increase in pension expenditure. Figure 8.5 depicts the trends between pension and life expectancy

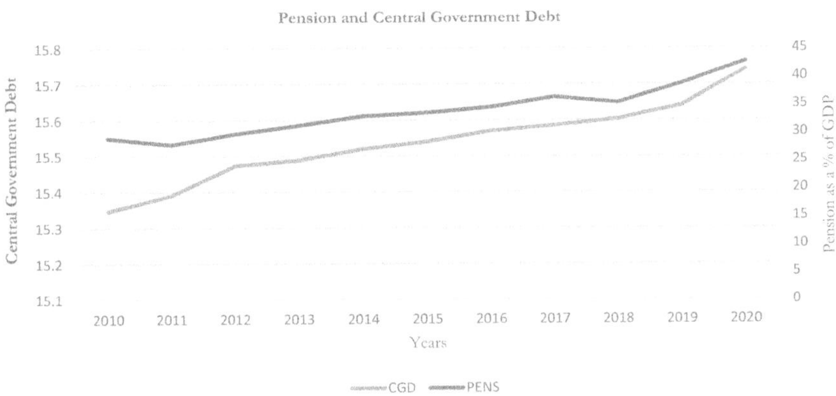


Figure 8.4: Trends between pension and central government debt in OECD countries

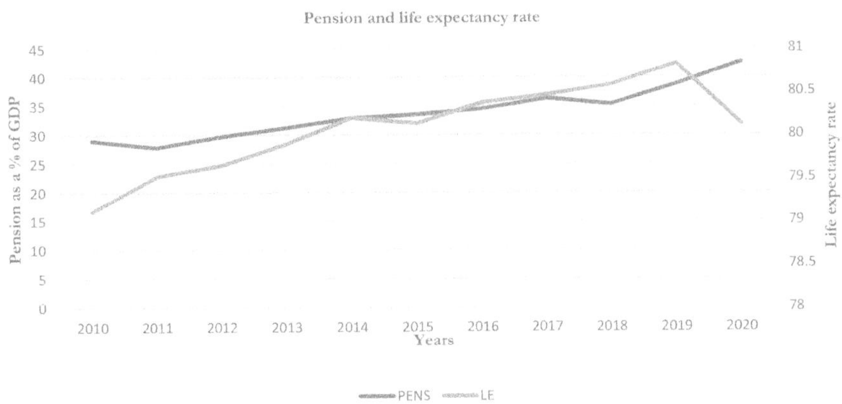


Figure 8.5: Trends between pension and life expectancy in OECD countries

and it looks like there occurs a positive association between them. But after the 2019 period, the pension increased, and life expectancy tremendously decreases because of covid-19 pandemic.

This also can impact government spending. Floristeanu (2013) in his paper found that life expectancy affects pension spending. With this background, the present paper examines the following hypothesis.

> $H_0$: There is no association between central government debt,life expectancy,& Pension.

> $H_1$: There is a association between central government debt,life expectancy,& Pension.

## Framework for Data and Models

We investigate the impact of socioeconomic issues on the pension scheme. The analysis made use of the OECD library and the World Bank database of all OECD countries. The data covers all 38 OECD countries shown in Table 8.A1 (see the Appendix section of this chapter). The variables employed were relevant to the specific research and were selected based on the body of existing literature. In our model, Pension is the dependent variable whereas factors like GDP per capita, institutional quality index, Central Government debt, life expectancy, and expenditure on health sector are the explanatory variables. The variables per capita GDP and Central Government Debt are taken in natural logarithm form however, the other variables are in their original form. Further, the description of the variables is represented in Table 8.1.

The standard deviation in Table 8.2 reveals the high volatility of all the variables under study. Additionally, between the minimum and maximum values of the variables have a substantial difference is seen. To check the normality of the distributed variables we have to test the Kurtosis and Skewness, and other semo-parametric test known as Shapiro wilk and Shapiro francia test. The results shows they are Non-normal in the nature as shown in the probability value.

Table 8.1: Description of the variables

| Indicator Name | Label | Data Source |
|---|---|---|
| Pension as a % of GDP | PENS | OECD Pensions Statistics (2021) |
| GDP per capita | GDPPC | World Development Indicators (December 2022) |
| Health expenditure as a % of GDP | HELEXP | OECD Health expenditure and financing (2021) |
| Institutional Quality Index* | IQI | Computed from the Worldwide Governance indicator (WGI) project (World bank, 2022) |
| Central Government debt as a % of GDP | CGD | World Development Indicators (December 2022) |
| Life Expectancy | LE | World Development Indicators (December 2022) |

**Source:** Authors' compilation from different sources as mentioned in the table. **Note:** *Computed index (through PCA) of six Worldwide Governance Indicators: (1) Voice and Accountability, (2) Political Stability and Absence of Violence/Terrorism, (3) Government Effectiveness, (4) Regulatory Quality, (5) Rule of law and (6) Control of Corruption.

Table 8.2: Descriptive statistics

| Variable | PENS | GDPPC | LE | HELEXP | IQI | CGD |
|---|---|---|---|---|---|---|
| Mean | 33.83 | 11.71 | 80.08 | 8.69 | 0.00 | 27.91 |
| Std.Dev. | 43.60 | 2.18 | 2.76 | 2.24 | 1.00 | 3.18 |
| Minimum | 0.02 | 9.20 | 70.13 | 4.12 | −1.65 | 20.72 |
| Maximum | 212.74 | 17.39 | 84.62 | 16.84 | 1.77 | 34.69 |
| Shapiro-Wilk Prob. | 0.00 | 0.00 | 0.00 | 0.00 | 0.00 | 0.00 |
| Shapiro-francia Prob | 0.00 | 0.00 | 0.00 | 0.00 | 0.00 | 0.00 |
| Skewness Normality Prob. | 0.00 | 0.00 | 0.00 | 0.00 | 0.04 | 0.00 |
| Kurtosis Normality Prob. | 0.00 | 0.22 | 0.53 | 0.00 | 0.00 | 0.72 |
| Obs | 418.00 | 418.00 | 418.00 | 418.00 | 418.00 | 418.00 |

**Source:** Authors' calculation.

Additionally, estimates for the empirical models are based on the regression equation shown below.

$$PENS = f\left(GDPPC, IQI, LE, HELEXP, CGD\right) \quad (8.1)$$

$$\begin{aligned} PENS_{it} = {} & \alpha_0 + \alpha_1 GDPPC_{it} + \alpha_2 IQI_{it} + \alpha_3 LE_{it} \\ & + \alpha_4 HELEXP_{it} + \alpha_5 CGD_{it} + \varepsilon_{it} \end{aligned} \quad (8.2)$$

Where, i = 1, 2 …, N (for the country) and t = 1, 2 …, t (for the number of years). Equation (8.1) shows the functional relationship of GDP per capita (GDPPC) and Institutional Quality Index (IQI) on pension expenditure with other explanatory and control variables.

## Econometric Method

Before empirical estimations, a variety of diagnostic tests are performed to determine the most effective techniques. We used non-parametric estimating methods because our dataset is not normally distributed in nature. To do an empirical analysis, the diagnostic tests for autocorrelation, multicollinearity, heteroscedasticity, and cross-sectional dependency are employed. According to the Wooldridge autocorrelation test (Wooldridge, 2010), the data collection has an issue of autocorrelation. The Wald test for group-wise heteroscedasticity demonstrates that there are heteroscedasticity issues in the error terms. Both the Hausman (Hausman, 1978) and the Breusch-Pagan LM (Breusch & Pagan, 1980) tests demonstrate that the model has a fixed effect (see Table 8.A2 in the Appendix). Besides, the Cross-sectional dependence (CSD) tests (Pesaran, 2021; Friedman, 1937; Frees, 1995) validate the error term's cross-sectional dependency (see Table 8.4). According to the results of the variance inflation factor (VIF) approach, there are no major multicollinearity issues among the variables (see Table 8.3).

The issue like heteroscedasticity, autocorrelation, and cross-sectional dependencies are addressed using the fixed effect (FE) model with Driscoll-Kraay standard errors (Driscoll & Kraay, 1998). Furthermore, our sample size is limited to 38 countries (n) and ten-time periods (t).The FE model with Driscoll-Kray standard errors is used to solve the discussed issues. Moreover, following (Hahn, 1995) this study has also employed the Bootstrapped Panel Quantile Regression (BPQR) method to study the effect of the above-mentioned explanatory variables on the pension system. This strategy is useful when studying with a small sample size because it reduces the impact of outliers in the dataset (Nikitina et al., 2019). The Westerlund panel cointegration test (Westerlund & Edgerton, 2007) is used to examine the long-term cointegration among the series. The findings indicate that the variables are cointegrated (see Table 8.4).

The following stages are involved in the estimate of the FE model with Driscoll Kraay standards errors (Hoechle, 2007). First, equation (8.2), as mentioned below transforms each variable in the model.

$$z_{it}^{\sim} = z_{it} - z_i^{-} + z_t^{ii} \tag{8.3}$$

Where

$$z_i^{-} = T_i^{-1} \sum_{t=t_{i1}}^{t_i} z_{it}$$

Further, the OLS estimation below (Equation 8.4) gives within the estimation.

$$y_{it}^{\sim} = x_{it}^{\sim'} \theta + \varepsilon_{it}^{\sim} \tag{8.4}$$

The pooled OLS estimation with Driscoll-Kraay standards errors is used to estimate the converted model of equation (8.4).

Furthermore, the BPQR model, which provides three distinct quantiles of the pension expenditure, is being used to test the consistency of the results. Three quantiles are chosen so that the 25th quantile represents the lower expenditure in pension (PENS) countries, the 50th quantile represents the medium expenditure in pension countries, and the 75th quantile represents the highest expenditure in pension countries. When BPQR is used for estimation and the dataset has a non-normal distribution, quantile regression is helpful since it mitigates the effects of group-wise heteroscedasticity in the residuals.

## Results and Discussion

This study has applied various pre-estimation tests and their results are discussed as follows. The results of the variance inflation factor (VIF) are reported in Table 8.3 shows that the average VIF score of the variables is less than two. Multicollinearity issues are discovered as a warning when the VIF number surpasses 10 (Marquardt, 1970; Mason et al., 1989; Hair et al., 1995).This implies that multicollinearity is not causing any significant problems in our study.

After multicollinearity, the results of cross-sectional dependence tests such as Pesaran (2021), Friedman (1937), and Frees (1995) are reported in Table 8.4, and shows the significant test statistics which rule out the null hypothesis of cross-sectional independence. As a result, we can say there is

Table 8.3. Multi-collinearity statistics

| **Variable** | **VIF** | **1/VIF** |
|---|---|---|
| HELEXP | 1.49 | 0.61 |
| CGD | 1.09 | 0.66 |
| LE | 1.63 | 0.70 |
| GDPPC | 1.07 | 0.91 |
| IQI | 1.42 | 0.93 |
| Mean VIF | 1.34 | |

**Source:** Authors calculation.

Table 8.4: Results of panel unit root tests (CADF and CIPS) & cross section dependence tests

| **Variable** | **CADF** | | **CIPS** | |
|---|---|---|---|---|
| | I (0) | I (1) | I (0) | I (1) |
| PENS | −1.983*** | −2.092** | −2.257 | −3.036* |
| GDPPC | −2.358* | −2.063** | −2.144 | −2.817** |
| LE | −1.539 | −2.093** | −2.996* | −3.271* |
| HELEXP | −1.243 | −1.704 | −2.509 | −2.836** |
| CGD | −0.032 | −0.171 | −0.838 | −0.968 |
| IQI | −2.579* | −2.372* | −2.067 | −3.364* |
| **Cross-section dependence test (CD)** | | | | |
| **Tests** | **Statistic** | | **P-Value** | |
| Breusch-Pagan LM | 3,232.83 | | 0.00 | |
| Pesaran Scaled LM | 67.62 | | 0.00 | |
| Pesaran CD | 37.50 | | 0.00 | |
| **Westerlund panel cointegration test** | | | | |
| **Statistic** | p-value | | | |
| 1.7861 | 0.0370 | | | |

**Source:** Authors calculation.

cross sectional dependence among the OECD nations. Our research shows that the shock in one country may spread to other nations. Following that, we use a second-generation panel unit root test, such as CADF or CIPS, to determine whether the variable is stationary. According to the findings, most of the variables are either level I (0) or first order I (1) stationary (see Table 8.4). Afterends, the Westerlund and Edgerton (2007) cointegration test is used and the results informed in Table 8.4 indicate that there is long-run cointegration among the series.

For the regression analysis, the paper uses the FE model with Driscoll and Kraay standard error as it solves the issues discussed above. The results of the estimation of equations (8.1) using the FE model with Driscoll and Kraay standard errors are displayed in Table 8.5, which shows that per capita

Table 8.5: FE model with Driscoll-Karry standards errors

| Dependent Variable: PENS | Coefficient | std. err. | t | P>t |
|---|---|---|---|---|
| Independent variables | | | | |
| GDPPC | 11.95 | 5.72 | 2.09 | 0.04 |
| HELEXP | 3.85 | 0.70 | 5.5 | 0.00 |
| CGD | 9.20 | 1.75 | 5.25 | 0.00 |
| IQI | 2.60 | 0.94 | 2.75 | 0.00 |
| LE | −1.27 | 0.91 | −1.39 | 0.17 |
| _cons | −181.08 | 93.66 | −1.93 | 0.06 |

**Source:** Authors calculation.

GDP positively impacts the pension expenditure. It shows that rich nations can spend more on pensions. Similarly, health expenditure and institutional quality index positively influence the pension spending for the OECD nations. There is also positive relationship between the central government debt and pension spending. However, life expectancy rate variable found to be statistically not significant.

As discussed, to check the consistency of the FE model with Driscoll and Kraay standard errors we have applied the Bootstrapped Panel Quantile Regression in this study (BPQR) to assess the effect of the mentioned explanatory variables on pensions expenditure across three quantiles (low, medium, and high) shown in Table 8.6. Further, Figure 8.6 explains the panel quantile plots. The pension (PENS) at various conditional quantiles is used as the dependent variable when compared to the independent variables GDP per capita (GDPPC), institutional quality index (IQI), health expenditure (HELEXP), central government debt (CGD), and life expectancy rate (LE). Using quantile regression, we can investigate effects that are potentially diverse and not mean-dependent, it also gives robust results when there are few cross-sections.

It is found that increasing GDP per capita causes a country's low 25th quantile pension amount to rise. The institutional quality index (IQI) and health expenditure (HELEXP) have also been identified as having a significant and positive relationship with pensions. In the lowest quantile, life expectancy (LE) has a negative impact on the pension. The 25th quantile

Table 8.6: Bootstrapped panel quantile regressions

| **Dependent Variables: Pension** | | | |
|---|---|---|---|
| **Variables** | **Lower Quantile (25)** | **Medium Quantile (50)** | **Higher Quantile (75)** |
| GDPPC | 1.36* | 2.75* | 4.50** |
| HELEXP | 0.96* | 6.12* | 4.64* |
| CGD | −0.03 | −0.15 | 0.28 |
| IQI | 0.82* | 5.83* | 7.19* |
| LE | −0.64* | −1.95* | 3.60* |

**Note:** p values in parentheses; *, **, and *** are statistically significant at 1 %, 5 %, and 10 % levels, respectively.

results are comparable to the 50th quantile results. However, the results differ at a higher quantile (75th quantile). The LE has a significant and positive influence on pensions in the 75th quantile however, it is negative and significant for the lower and middle quantile countries. Similarly, the result varies for the central government debt between different quantiles of pension spending economies.

The institutional quality index (IQI) and GDP per capita both have a positive and significant impact on the social welfare (pension spending) of the nation across all the quantiles. Furthermore, the health expenditure coefficients are positive and significant, indicating that rising health spending raises pensions across all the quantiles. According to the findings, central government debt has a adverse and insignificant impact on pensions in the lower and middle quantiles. This is consistent with the hypothesis, though the effect is insignificant and positive at a higher quantile. The same results have been observed in terms of life expectancy which shows a negative and statistically significant impact on pensions in the lower and middle quantiles, but a positive and significant impact in the upper quantile.

The panel quantile plots for the estimated variables are shown in Figure 8.6 for quantiles up to 1 and 0.2 intervals. The outcome of quantile plots shows that the influence of the institutional quality index, per capita GDP, and health spending on pension is positive across all quantiles. Additionally, the effect of central government debt on pensions is negative up to 0.6 quantiles, positive up to 0.9 quantiles, and then negative again beyond that. This demonstrates that while the central government debt (CGD) has a adverse impact on pensions in nations with low and high pension spending economies. Similar outcomes are observed when examining life expectancy (LE), up to the 0.6th quantile, life expectancy has a negative impact on pensions, after which it has a favorable impact. Therefore, the illustration demonstrated how life expectancy has a favorable impact on pensions in countries where pension expenditure is high.

This study examined the impact of per capita GDP, institutional quality index, health expenditure, central government debt, and life expectancy on the pension system in the OECD countries by using annual data from 2010 to 2020. The fixed effect Driscoll standard error model examines the relationship between pension and its determinants. The Husman test supports using the fixed effect Driscoll standard error model. Further, to check the robustness of the estimated model the bootstrap panel quantile regression (BPQR) is applied which analyzed the relationship between the variables at different quantiles. The quantiles are based on the pension spending by the countries. The countries which spend the lower amount on pension as a percentage of GDP are considered a lower quantile group of countries, similarly, the countries who spend a higher amount are considered under the upper quantile, and countries in between are median quantile.

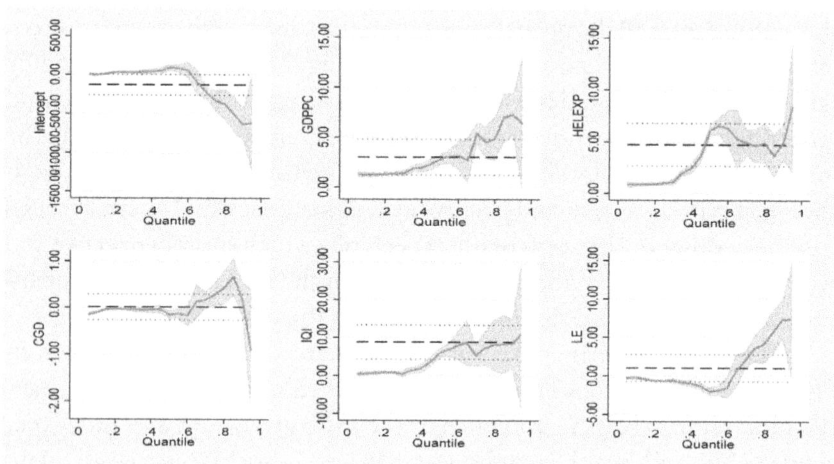


Figure 8.6: Panel quantile plots

## Conclusion and Policy Implications

The finding indicates that the per capita GDP, health expenditure, institutional quality index, and government debt are positive and significantly associated with the pension system in the OECD countries.

Furthermore, the Bootstrapped panel quantile regression results also support the results of the fixed effect model except for the few variables. Here the per capita GDP, institutional quality index, and health expenditure are positively and significantly affected to the pension expenditure in all the quantiles. At the same time, the relationship between central government debt and pension expenditure is negative for the lower and middle pension expenditure countries and positive for higher quantile countries. However the coefficient are not statistically significant. Additionally, life expectancy is negatively affected to lower and middle-pension expenditure countries and positively affected to the high pension expenditure countries significantly. Worldwide, governments are making greater efforts to increase pensions and other forms of social assistance. The pension system implementation is based on the institution's score on the institutional quality index. The complex relationship between the pension system and other economic factors makes it difficult to analyze the direct and indirect impact of each explanatory variable. This rejects the null hypothesis 1–3 and confirms the positive relationship between them. In contrast, life expectancy does not significantly affect the pension system. This partly accepts hypothesis 4 and indicates the no significant relationship between life expectancy and pension. But surprisingly, it is positively related to the higher quantile economies and negatively related to other quantiles.

The findings from the Panel quantile plots and bootstrapped panel quantile regressions help in elucidating additional details about the nations with low, medium, and high pension expenditure. All quantiles of national pension systems benefit from increased per capita GDP. In contrast, life expectancy is negatively and significantly affects the pension up to 0.6 quantiles, but after that point, they positively affect the pension. Furthermore, the institutional quality index and health expenditure improve pensions across all quantiles of economies. Additionally, the panel quantile plots show that life expectancy positively impacts the pension in the higher quantile nations above the 0.6th quantile. In general, the strength of a country's retirement system is determined by several factors, including its GDP per capita, institutional quality index, and health spending.

One of the most significant aspects of OECD nations' efforts to boost social welfare is their payment systems (European Commission, 2012). Several factors, such as GDP per capita, the institutional quality index, health expenditure, and the character of the central government, all contribute positively to the social welfare of a country, while life expectancy has an insignificant effect. Therefore, life expectancy has little bearing on the welfare system. Thus, the stakeholders of all OECD countries and policymakers should take such initiative and frame such policies to improve the pension systems of all OECD countries to achieve more social welfare. This paper is limited to OECD countries due to lack of data for the selected variables under study.

## Appendix

Table 8.A1: List of sample countries

| | | | | |
|---|---|---|---|---|
| Australia | Denmark | Ireland | Mexico | Spain |
| Austria | Estonia | Israel | Netherlands | Sweden |
| Belgium | Finland | Italy | New Zealand | Switzerland |
| Canada | France | Japan | Norway | Türkiye |
| Chile | Germany | Korea | Poland | United Kingdom |
| Colombia | Greece | Latvia | Portugal | United States |
| Costa Rica | Hungary | Lithuania | Slovak Republic | |
| Czech Republic | Iceland | Luxembourg | Slovenia | |

Table 8.A2: Results of diagnostics tests

| **Tests** | **Test statics** | **P value** |
|---|---|---|
| Hausman test | 16.56 | 0.00 |
| Breusch pagan LM test | 1,850.92 | 0.00 |
| Modified Wald test for group-wise heteroskedasticity | 390,000 | 0.00 |
| Wooldridge test for autocorrelation | 59.72 | 0.00 |

## Editors Note

The investigation into the determinants of pensions, supported by empirical evidence from OECD countries, holds significant implications for sustainability in retirement systems and social welfare policies. Pensions play a crucial role in ensuring financial security and well-being for retirees, particularly as populations age and life expectancies increase. By examining the factors influencing pension provision and adequacy across OECD countries, this research can offer valuable insights into designing sustainable pension systems that meet the needs of current and future generations. Understanding the drivers of pension outcomes, such as demographic trends, labor market dynamics, and policy interventions, can inform strategic decision-making processes aimed at enhancing the effectiveness, equity, and long-term viability of pension schemes. Moreover, by identifying best practices and policy innovations from OECD countries, policymakers can develop targeted reforms to address challenges related to pension adequacy, affordability, and fiscal sustainability. Therefore, this empirical investigation has the potential to inform policy debates and reforms aimed at promoting sustainable retirement income systems and ensuring financial security for individuals throughout their later years.

## References

Alon, I. (2014). *Social franchising.* London: Palgrave Pivot/ Springer. <https://link.springer.com/book/10.1057/9781137455840>.

An, Y., Zhao, K., & Zhou, R. (2016). Health spending and public pension: Evidence from panel data. *Applied Economics, 48*(11), 987–1004.

Azolibe, C. B. (2022). Panel data comparative analysis on the influence of population age structure on the size of government expenditure in Africa and Asia. *International Journal of Emerging Markets, 17*(2), 433–451.

Been, J., Caminada, K., Goudswaard, K., & Van Vliet, O. (2017). Public/private pension mix, income inequality and poverty among the elderly in Europe: An empirical analysis using new and revised OECD data. *Social Policy and Administration, 51*(7), 1079–1100.

Breusch, T. S., & Pagan, A. R. (1980). The Lagrange multiplier test and its applications to model specification in econometrics. *The Review of Economic Studies, 47*(1), 239–253.

Brown, A. J., & Fraikin, A. L. (2022). The old-age pension household replacement rate in Belgium. *The Journal of the Economics of Ageing, 23*, 100402.

Clark, R. L., Craig, L. A., & Wilson, J. W. (1999). The life and times of a public-sector pension plan before social security: The US Navy pension plan in the nineteenth century. In O. S. Mitchell & E. C. Hustead (Eds), *Pensions in the public sector* (pp. 1–1999). Philadelphia: University of Pennsylvania Press.

Driscoll, J. C., & Kraay, A. C. (1998). Consistent covariance matrix estimation with spatially dependent panel data. *Review of Economics and Statistics, 80*(4), 549–560.

European Commission. (2012). *An agenda for adequate, safe and sustainable pensions: White paper*, Publications Office of the European Union.

Firpo, S., Fortin, N. M., & Lemieux, T. (2009). Unconditional quantile regressions. *Econometrica, 77*(3), 953–973.

Floristeanu, E. (2013). Repercussions of economic and social factors on pension systems. *Procedia Economics and Finance, 6*, 627–633.

Frees, E. W. (1995). Assessing cross-sectional correlation in panel data. *Journal of Econometrics, 69*(2), 393–414.

Friedman, M. (1937). The use of ranks to avoid the assumption of normality implicit in the analysis of variance. *Journal of the American Statistical Association, 32*(200), 675–701.

Hahn, J. (1995). Bootstrapping quantile regression estimators. *Econometric Theory, 11*(1), 105–121.

Hair Jr J. F., Anderson, R. E., Tatham, R. L., & Black, W. C. (1995). *Multivariate data analysis* (3rd edn). New York, NY: Macmillan.

Hausman, J. A. (1978). Specification tests in econometrics. *Econometrica. Journal of the Econometric Society, 46*, 1251–1271.

Heer, B., & Trede, M. (2023). Age-specific entrepreneurship and PAYG: Public pensions in Germany. *Journal of Macroeconomics, 75*, 103488, 281–312.

Hoechle, D. (2007). Robust standard errors for panel regression with cross-sectional dependence. *The Stata Journal, 7*(3), 281–312.

Holzner, M., Jestl, S., & Pichler, D. (2022). Public and private pension systems and macroeconomic volatility in OECD countries. *Scottish Journal of Political Economy, 69*(2), 131–168.

Hwang, I., & Lee, T. J. (2022). Health improvements of older adults based on benefit duration: Lessons from Korean social pension policies. *Social Science & Medicine, 315*, 115514.

International Human Rights Law. (2019). (RAIO). Available at: <https://www.uscis.gov/sites/default/files/document/foia/International_Human_Rights_Law_RAIO_Lesson_Plan.pdf>

Joubert, C., & Todd, P. E. (2022). Gender pension gaps in a private retirement accounts system: A dynamic model of household labor supply and savings. *Journal of Econometrics, 243*(1–2), 105337. <https://doi.org/10.1016/j.jeconom.2022.08.005>.

Juergens, F. (2020). *HelpAge International: Why social pension?*. Available at: <https://docslib.org/doc/4585568/why-social-pensions-achieving-income-security-for-all-in-older-age-about-helpage.>

Kaganovich, M., & Zilcha, I. (2008). *Alternative pension systems and growth*, Working paper, Center for Economic Studies and Ifo Institute for Economic Research, University of Munich, December 8.

Kalish, D. W., & Aman, T. (1998). *Ageing working papers*, Working paper, Organization for Economic Co-operation and Development.

Kang, J. Y., Park, S., & Ahn, S. (2022). The effect of social pension on consumption among older adults in Korea. *The Journal of the Economics of Ageing, 22*, 100364.

Kuitto, K., Madia, J. E., & Podesta, F. (2023). Public pension generosity and old-age poverty in OECD countries. *Journal of Social Policy, 52*(2), 256–275.

Li, H., Li, H., Cao, A., & Guo, L. (2022). Does attending in social pension program promotes household energy transition? Evidence from ethnical minority regions of rural China. *Energy for Sustainable Development, 70*, 361–370.

Marquaridt, D. W. (1970). Generalized inverse, ridge regression, baised linear regression, and non-linear estimation. *Techno Metrics, 12*(3), 591–612.

Mason, R. L., Gunst, R. F., & Hess, J. L. (1989). *Statistic design and analysis of experiments: With applications to engineering and science* (pp. 568–606). New York, NY: Wiley.

Mazonga Mfoutou, B. O., & Danquah, R. (2023). Evaluating the efficiency of defined benefit pension plans in a developing market. *International Journal of Emerging Markets, 11*, 4116–4131.

Mostert, C. M., Mackay, D., Awiti, A., Kumar, M., & Merali, Z. (2022). Does social pension buy improved mental health and mortality outcomes for senior citizens? Evidence from South Africa's 2008 pension reform. *Preventive Medicine Reports, 30*, 102026.

Nikitina, L., Paid, R., & Furuoka, F. (2019). Using bootstrapped quantile regression analysis for small sample research in applied linguistics: Some methodological consideration. *PloS One, 14*(1), 0210668.

OECD. (2021). *Pensions at a Glance 2021: OECD and G20 indicators.* Paris: OECD Publishing.

Penas, B. M. C., Brinker, L. V., Eisenberg, J., & Korn, R. (2023). Managing reputational risk in the decumulation phase of a pension fund. *Insurance: Mathematics and Economics, 109*, 52–68.

Pesaran, M. H. (2021). General diagnostic tests for cross-sectional dependence in panels. *Empirical Economics, 60*(1), 13–50.

Romp, W., & Beetsma, R. (2022). OECD pension reform: The role of demographic trends and the business cycle. *European Journal of Political Economy, 77*, 102280.

Wang, W., Shi, H., & Li, Q. (2023). Pension gap between the Chinese public and nonpublic sectors: Evidence in the context of the integration of dual-track pension schemes. *International Review of Economics & Finance, 85*, 664–688.

Westerlund, J., & Edgerton, D. L. (2007). A panel bootstrap cointegration test. *Economics Letters, 97*(3), 185–190.

Wooldridge, J. M. (2010). *Econometric analysis of cross-section and panel data.* London: MIT Press.

DR. SUSHMA AND DR. ANURAG SHUKLA

# 9 Associability between Overconfidence Bias and Financial Risk Tolerance with Reference to North India

ABSTRACT

According to studies worldwide approximately 93 % of population poll their skills as above average, which can't be significantly true always. This above average rating effect is the tendency of individuals to assume themselves as above average on different characteristic abilities that are both ambiguous and socially desirable (Myers, 2002). It elucidates that people sometimes are overconfident about their abilities. In a realistic world, a confident investor is one who is more optimistic towards their abilities than knowledge. Overconfidence can be detrimental to the investors stock picking ability.

Studies show overconfident investors tend to do more trades and have more moderately tender counterparts but more trades don't necessarily mean better results. Generally, the investor who conducts more trade yield lower average returns. There is a very fine line between confidence and overconfidence; but this fine line can actually make larger differences. Although rating one as above average has been documented for a variety of different characteristics and classroom demonstration of this effect already, exist (Bolt, 1998). However there is a lack of specific results to check and explain the reasons and parameters of developing this psychology among investors. The paper here tries to identify this overconfidence bias among investors and its association with their financial risk tolerance capacity.

Keywords: Risk, overconfidence bias, risk-tolerance scale, psychology, investment decision

## Introduction

Generally, investors' investing decisions may also be influenced by illogical factors relating to psychology, which are frequently referred to as financial behavior. The two basic categories of financial behavior are cognitive psychology and the limit of arbitrage (Ritter, 2003). The focus of cognitive psychology is on thought processes. Numerous pieces of evidence demonstrate that cognitive bias leads to improper decisions. Every investor must choose between projected return and risk when making an investment decision. Consequently, an investor's view of risk may have an impact on his decision to make an investment (Nofsinger, 2017; Pompian, 2012).

The way investors perceive risk is one bias that has an impact on their choices. A person with a high-risk tolerance might view a low-risk investment option as carrying a larger risk, whereas a person with a low-risk tolerance might view a high-risk asset as carrying a lower risk. Because of this, an investor with a high-risk tolerance will typically opt to allocate funds to low-risk assets, whilst an investor with a low-risk tolerance will typically be more ready to commit funds to high-risk assets (Broihanne, Merli, & Roger, 2014; Weber & Milliman, 1997).

The degree of risk tolerance can also have an impact on an investor's investment choices. Investors' attitudes towards risk are displayed while evaluating a risk. In this situation, the risk is the unpredictability of the investment returns. Investors who are courageous in allocating funds to high-risk assets are typically those who are willing to accept or tolerate risk, and the opposite is also true (Corter & Chen, 2006).

Another behavioral bias that might affect decision-making is overconfidence. Overconfidence is a conviction that a person's judgment, cognitive aptitude, rational thought, and intellectuality in which he exaggerates his capacity for prediction and the accuracy of his own information (Pompian, 2012). Overconfident investors frequently underestimate risk, which can result in an unfavorable asset allocation (Dittrich, Güth, & Maciejovsky, 2005). As a result, the investor will typically allocate money to high-risk assets like stocks and real estate, whereas an investor with less overconfidence will put more of his money into low-risk assets.

## Review Literature

According to Nevins (2004), overconfidence is the tendency of investors to overestimate their capacity to predict market events. As a result, investors frequently take risks without realizing corresponding rewards. This article examines how combining conventional financial theory with behavioral finance insights can improve wealth management. Focus topics include risk measurement, risk profiling, and behavioral bias control. The author highlights the significance of taking investor preferences and goals into account when it comes to risk measurement, and he suggests various effective techniques for doing so. Additionally, he criticizes the practice of common risk profiling and contends that, rather than estimating an investor's overall risk tolerance, separate risk tolerance estimates for distinct goals should be used. He makes the point that the total portfolio framework of traditional finance is incompatible with investors' mental accounting tendencies. A better outcome can be achieved by connecting certain approaches to a specific target or goal. The author illustrates a method for putting his theories into practice using examples while taking into account the challenges of investing for a certain planning horizon and investing to cover current lifestyle needs. According to the research, this will establish credibility between the client's perspective and the practitioner's investment philosophies.

According to Odean (1998), overconfident investors don't necessarily manage and control risk inadvertently since they regularly trade in the market and acquire information from various sources. People frequently display excessive confidence. Overconfidence has an impact on financial markets. The forecast trading volume, market depth, and the anticipated utility of overconfident traders all increase with overconfidence. Those with excessive confidence have an impact on price quality and volatility. Due to overconfident traders, markets may overreact to news from rational traders. Additionally, markets overreact to immediate, anecdotal, and irrelevant material while underreacting to abstract, statistical, and extremely important information.

Making decisions is a crucial step in the process of choosing an alternative option from a situation that has favorable effects for people or investors, according to Statman et al. (2006) Making more money for investors is one of the fundamental goals of investing. Investor decision-making is significantly influenced by the informational structure and market characteristics. The excessive trade volume observed may be due to investors' overconfidence in their ability to value and trade securities. When bias self-attribution is applied, the degree of investor overconfidence and, as a result, trading volume, change with prior performance. The studies mainly look at the amount of trading estimates of accurate overconfidence models and find out how much shares are traded.

Institutional investors were generally regarded as being the least overconfident, but investment advisors were generally regarded as being the most overconfident, per (Menkhoff et al., 2013) findings. This study presents a comprehensive online experiment on stock market overconfidence. It often consists of three different sorts of members, including institutional investors, investors, and investment advisors, on a sizable web platform that tracks market sentiment. Due to their registration, certain socioeconomic aspects can be taken into consideration in our research. The three investing groups were shown to have consistently different levels of overconfidence. The level of overconfidence is also significantly influenced by age and investment experience, which unexpectedly has the opposite effect.

Graham et al. (2009) studied high-confidence investors that frequently invest in foreign assets. The main objective of this study is to determine whether the "competence effect" has an influence on the frequency of trading. People are more willing to trust their own judgments when they feel knowledgeable or skillful. According to the research, most investors have a strong sense of competence, they trade more frequently, and their portfolios are more widely diversified. It also concludes that male investors, those with higher incomes or more education, and those who view themselves as informed investors are more likely than female investors, those with lower incomes or less education.

Women are less confident than men in their understanding and awareness of the Indian stock market, according to research by Prosad et al. (2015). The purpose of this study is to examine the presence of behavioral biases among Indian investors, including overconfidence, excessive optimism (pessimism), herd behavior, and the disposition effect. It next examines how demographics and investor intelligence affect biases. Finally, it pinpoints the prejudices that are most pervasive in India. Given that each bias is associated with a particular set of investor characteristics, overconfidence emerges as the most pertinent bias in the Indian context. The study offered a unique viewpoint on the investor profile that matches to each problematic prejudice. It not only updates knowledge on behavioral biases but also pinpoints which prejudice affects Indian society the most.

According to Khan et al. (2017) research, overconfidence biases always have a significant and advantageous impact on investment results. The purpose of the study was to ascertain how Khyber Pakhtunkhwa has perceived trading return on the Pakistan Stock Exchange affected investors' perceptions of behavioral biases. All investors in Khyber Pakhtunkhwa who made investments in First National Equities Limited and TAURUS Securities made up the study's population. The results show that the majority of managers and academics think that behavior biases are the most crucial tool for gaining an advantage in trading and that investors and organizations need to be aware of investor and market behavior in order to have high levels of success at the stock exchange.

Similar research was undertaken by (Prasad et al.) with a focus on overconfidence bias, additional optimism, disposition effect, and herding behavior bias in order to examine biased behavior among several Indian stock market investors. The most potent inhibitors of these four biases were found to be age, occupation, and the frequency of human trafficking. In terms of their understanding of the Indian stock market, male investors are more assured than female investors are. Women are less optimistic and gloomy than men because they think the price of gold will increase soon; as a result, they prefer to invest more in gold rather than the Indian stock market.

In Chennai, India, (Isidore & Christie's, 2019) investigated the relationship between investors' annual income and behavioral biases. This study found that investors with higher income levels are less likely to be prejudiced than those with lower income levels. High-income investors have a modest chance of having overconfidence biases; whereas low-income investors primarily exhibit loss aversion, regret version, etc.

## Objectives

1. To investigate the level of financial risk tolerance among North Indian investors using GL-RTS scale.
2. To study overconfidence bias and its relativeness with demographic profile of investors.
3. To analyze the association between overconfidence bias and Financial Risk Tolerance among north Indian investors.

## Research Methodology

**Research Design:** The research design here is exploratory in nature where the statistical techniques like correlation; regression analysis is used to interpret the relationship among variables.
**Sample Area and Size:** The sample here consists of investors from top five North Indian states based on number of investors (registered clients) as per data published at BSE. It consists Uttar Pradesh, Rajasthan, Madhya Pradesh, Delhi and Haryana. The data is targeted to be collected from 181 investors in the same ratio.
**Sampling Technique:** Sampling technique used here is stratified random, where the investors from five different states of north India is taken for sampling.

Table 9.1: Sample size estimation

| S. No. | State Selected | Number of Registered Clients (As on April 2023) | Ratio (%age) |
|---|---|---|---|
| 1 | Uttar Pradesh | 12,386,751 | 9.6 |
| 2 | Rajasthan | 7,448,735 | 5.7 |
| 3 | Madhya Pradesh | 6,457,140 | 5.2 |
| 4 | Delhi | 6,041,535 | 4.6 |
| 5 | Haryana | 4,211,519 | 3.2 |
| | | 128,805,392 (Total in India) | |

## Hypothesis Used:

($H_0$): There is no relationship between Overconfidence bias and financial risk tolerance.

($H_1$): There is a relationship between Overconfidence bias and financial risk tolerance.

**Data Collection:** The data for this study is collected from primary source in the form of structured questionnaire and their responses, which is divided in two parts.
*Part A* is associated with calculation of Financial Risk Tolerance where the 13 items scale given by John E. Grable and Lytton, is used; that categorized investor in five categories, according to their response score.

1. High risk tolerant
2. Moderate to high risk tolerant
3. Moderate risk tolerant
4. Low to moderate risk tolerant
5. Low risk tolerant

*Part B* is related to calculating the *overconfidence biasness* among investors where five-point Likert scale is used. In this scale one point represents strongly disagree and five points represent strongly agree, as given in Table 9.2.

Table 9.2: Overconfidence bias estimation questions

| No. | Questions | Strongly Disagree (1) | Disagree (2) | Neutral (3) | Agree (4) | Strongly Agree (5) |
|---|---|---|---|---|---|---|
| 1 | You trust your capabilities fully to start and manage new investment portfolios. | | | | | |
| 2 | Your previous successful investments have boosted your level of confidence. | | | | | |
| 3 | You feel that your earlier investment decisions often proven you correct. | | | | | |
| 4 | You never hesitate to offer any financial advice to your well-wishers. | | | | | |
| 5 | You believe that your investment options has always outperformed in past. | | | | | |
| 6 | You rely on your analytical skills to evaluate all your investment decisions yourself. | | | | | |
| 7 | You feel that your specific investment decision skills provide you an extra edge to earn more profits. | | | | | |
| 8 | You believe that your returns are always higher than other (same risk bearing option's) market returns. | | | | | |

## Data Analysis and Interpretation

### *Calculation of Level of Financial Risk Tolerance*

Based on 13 items GL-RTS given by Grable and Lytton, the respective scores are summed up. It shows that:

1. The score of 18 or below represents Low risk tolerance
2. The score of 19–22 shows Below-average risk tolerance
3. The score of 23–28 represents Average or moderate risk tolerance
4. The score of 29v32 infers Above-average risk tolerance
5. The score of 33 and above represents High risk tolerance

The frequency and count of investors is shown in Table 9.3 below.

The above data is also represented in Figure 9.1 below that shows the highest percentage of 46.4 % investors; belong to the category of below average risk tolerant and lowest percentage of 1.7 % investors; belong to above average risk tolerance level.

Table 9.3: Categories of investor as per financial risk tolerance level Investor risk tolerance

| | **Frequency** | **Percentage** | **Valid Percentage** | **Cumulative Percentage** |
|---|---|---|---|---|
| Low risk tolerance | 74 | 40.9 | 40.9 | 40.9 |
| Below average risk tolerance | 84 | 46.4 | 46.4 | 87.3 |
| Moderate risk tolerance | 16 | 8.8 | 8.8 | 96.1 |
| Above average risk tolerance | 4 | 2.2 | 2.2 | 98.3 |
| High risk tolerance | 3 | 1.7 | 1.7 | 100.0 |
| Total | 181 | 100.0 | 100.0 | |

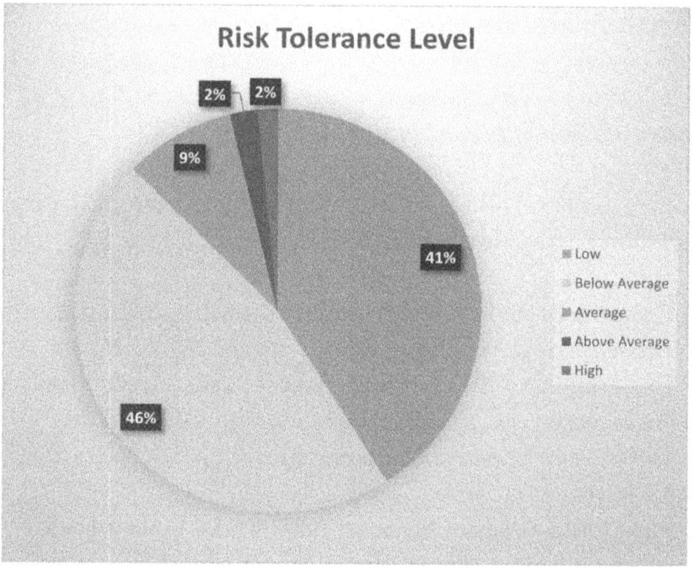


Figure 9.1: Graphical presentation of financial risk tolerance level

## Overconfidence Bias and Its Relativeness with Demographic Profile of Investors

For this objective, the mean comparison techniques have been used. The results shown below in Table 9.4 explains that, in context of gender; overconfidence bias is found high in males as compared to female investors. Males are considered aggressive investors as compared to females.

Table 9.4: Overconfidence bias and gender

| Gender | Mean | N | Std. Deviation |
|---|---|---|---|
| Male | 2.32 | 37 | 1.292 |
| Female | 1.72 | 144 | 0.575 |
| Total | 1.84 | 181 | 0.811 |

Now, considering the age as demographic factor the results suggest that overconfidence bias is highest in youngsters of age group 26–35 years while lowest in the old age investors of the group 46–55 years. Table 9.5 below shows the same, where it is visible that young age investor are highly overconfident.

The next demographic factor considered is marital status of investors, where the results show that overconfidence bias is highest in those who are unmarried while the lowest in those investors who are single parent (separated). Table 9.6 below represents the data of marital status and overconfidence association. The reason may be the level of risk tolerance, which is highest in unmarried as they having lesser responsibilities as compared to those who are single parent.

Considering monthly income as next demographic factor the data shows that high overconfidence level is related to high income group. The highest is in those having monthly income of Rs. 80,001–1,00,000 and lowest in those having up to Rs. 30,000. The Table 9.7 below shows the different values of means associated with all income levels.

Table 9.5: Overconfidence bias and age

| Age | Mean | N | Std. Deviation |
|---|---|---|---|
| 18–25 years | 2.89 | 18 | 1.023 |
| 26–35 years | 2.92 | 12 | 1.379 |
| 46–55 years | 1.63 | 151 | 0.511 |
| Total | 1.84 | 181 | 0.811 |

Table 9.6: Overconfidence bias and marital status

| Marital | Mean | N | Std. Deviation |
|---|---|---|---|
| Unmarried | 3.36 | 11 | 0.924 |
| Married without children | 1.76 | 29 | 0.636 |
| Married with children | 1.74 | 136 | 0.720 |
| Widowed | 1.67 | 3 | 0.577 |
| Separated | 1.50 | 2 | 0.707 |
| Total | 1.84 | 181 | 0.811 |

Table 9.7: Overconfidence bias and income

| Income | Mean | N | Std. Deviation |
|---|---|---|---|
| Upto Rs. 30,000 | 1.60 | 40 | 0.496 |
| Rs. 30,001–Rs. 60,000 | 1.50 | 92 | 0.503 |
| Rs. 60,001–Rs. 80,000 | 3.31 | 16 | 0.479 |
| Rs. 80,001–1,00,000 | 4.00 | 7 | 0.000 |
| Nil | 1.92 | 26 | 0.272 |
| Total | 1.84 | 181 | 0.811 |

## Association between Overconfidence Bias and Financial Risk Tolerance

Finally in order to check the associability, correlation analysis is used; that interprets the degree and direction of association between two variables. Pearson's correlation coefficient is used here in order to check the research hypothesis shown below.

($H_0$): There is no relationship between Overconfidence bias and financial risk tolerance.

($H_1$): There is a relationship between Overconfidence bias and financial risk tolerance.

Coefficient value "r" as given in Table 9.8 shows that a significant relationship is found between overconfidence bias and financial risk tolerance ($p = 0.000$) which is less than level of significance resulting that null hypothesis is not accepted. Hence, research hypotheses is accepted that is, there is an association between overconfidence bias and risk tolerance. The degree of relationship; however, is found to be positive and strong.

The Figure 9.2 above shows the curve of positive coefficient, which infers that higher overconfidence bias, is found in the investors having high-risk tolerance, as they are confident about their investments and have hedging strategies to compensate against losses.

Table 9.8: Correlation coefficient between overconfidence bias and financial risk tolerance correlation

| | | overconf | Risk_tol |
|---|---|---|---|
| overconf | Pearson Correlation | 1 | 0.632** |
| | Sig. (2-tailed) | | 0.000 |
| | N | 181 | 181 |
| Risk_tol | Pearson Correlation | 0.632** | 1 |
| | Sig. (2-tailed) | 0.000 | |
| | N | 181 | 181 |

** Correlation is significant at the 0.01 level (2-tailed).

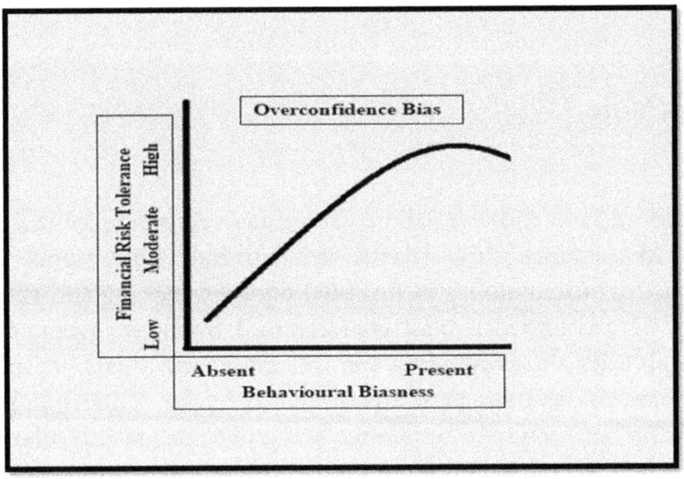


Figure 9.2: Correlation curve between overconfidence bias and financial risk tolerance

## Conclusion

Based on the results of the research it can be concluded that there is a strong positive relation between the overconfidence bias of investors and risk tolerance of investors. This means that investors who are overconfident in the financial market will be having a high level of risk tolerance. To prevent and reduce illogical decision-making, the investor community should be aware of behavioral bias. As a result, it must be taken into consideration when choosing investment assets. For investment advisors, it is expected that they can assist in learning about the characteristics of their clients, particularly those that are risk-related, to be aware of the potential for behavioral bias that their clients may experience, and to properly guide the clients when allocating their own funds.

## Editors Note

The exploration of the associability between overconfidence bias and financial risk tolerance with reference to North India holds significant implications for sustainability in financial decision-making and economic stability. Overconfidence bias, characterized by an exaggerated belief in one's abilities and judgments, can influence individuals' attitudes towards financial risk-taking and investment behavior. Understanding the relationship between overconfidence bias and financial risk tolerance in the context of North India can provide valuable insights into the factors driving investmentp decisions and market dynamics in the region. By identifying the prevalence and impact of overconfidence bias on financial risk tolerance, policymakers, regulators, and investors can develop targeted interventions aimed at promoting informed decision-making, risk awareness, and financial resilience. Moreover, by fostering a culture of risk management and investor education, stakeholders can contribute to the stability and long-term sustainability of financial markets in North India, ensuring their ability to support economic growth, wealth creation, and prosperity for individuals and communities. Therefore, this research has the potential to inform policy initiatives and industry practices aimed at promoting sustainable financial behavior and mitigating systemic risks in North India's financial markets.

## References

Broihanne, M.-H., Merli, M., & Roger, P. (2014). Overconfidence, risk perception and the risk-taking behavior of finance professionals. *Finance Research Letters, 11*(2), 64–73.

Corter, J. E., & Chen, Y.-J. (2006). Do investment risk tolerance attitudes predict portfolio risk? *Journal of Business and Psychology, 20*(3), 369.

Dittrich, D. A., Güth, W., & Maciejovsky, B. (2005). Overconfidence in investment decisions: An experimental approach. *The European Journal of Finance, 11*(6), 471–491.

Graham, J. R., Harvey, C. R., & Huang, H. (2009). Investor competence, trading frequency, and home bias. *Management Science, 55*(7), 1094–1106, *Journal of Wealth Management, 6*, no. 4 (2004), 8–23.

Khan, Y., Batool, S., & Khan, A. W. (2017). Impact of behavior biases on perceived trading return. *Journal of Business & Tourism, 3*(1), 63–73.

Menkhoff, L., Schmeling, M., & Schmidt, U. (2013). Overconfidence, experience, and professionalism: An experimental study. *Journal of Economic Behavior & Organization, 86*, 92–101.

Myers, D. G. (2002). *Social psychology* (7th edn). New York, NY: McGraw-Hill.

Nevins, D. (2004). Goals-based investing: Integrating traditional and behavioral finance. *The Journal of Wealth Management, 6*, 8–23.

Nofsinger, J. R. (2017). *The psychology of investing* (6th edn). New York, NY: Routledge.

Odean, T. (1998). Are investors reluctant to realize their losses?. *The Journal of Finance, 53*(5), 1775–1798.

Pompian, M. M. (2012). *Behavioral finance and investor types: Managing behavior to make better investment decisions*: New Jersey: Wiley.

Prosad, J. M., Kapoor, S., & Sengupta, J. (2015). Behavioral biases of Indian investors: A survey of Delhi-NCR region. *Qualitative Research in Financial Markets, 7*(3), 230–263.

Renu Isidore, R., & Christie, P. (2019). The relationship between the income and behavioural biases. *Journal of Economics, Finance and Administrative Science, 24*(47), 127–144. <https://doi.org/10.1108/JEFAS-10-2018-0111>.
Ritter, J. R. (2003). Behavioral finance. *Pacific-Basin Finance Journal, 11*(4), 429–437.
Statman, M., Thorley, S., & Vorkink, K. (2006). Investor overconfidence and trading volume. *The Review of Financial Studies, 19*(4), 1531–1565.
Weber, E. U., & Milliman, R. A. (1997). Perceived risk attitudes: Relating risk perception to risky choice. *Management Science, 43*(2), 123–144.

SHWETA LALWANI

# 10 Technology Enforcing Upskilling of Workforce for Creating Sustainable Organizations: A Conceptual Review of Trends and Transcending Role of HR

ABSTRACT
Use of artificial intelligence, Chabot's at workplace, business interactions in met averse are a few trends leading to digital revolution in performing the HR function. It has also shifted the role of HR professional from an advisory to a strategic one. The digital revolution has challenged the sustainability of the organizations by means of skill requirements. If the organizations fail to reskill or upskill the employees soon the viability of the organizations would be endangered. Hence, it is the need of hour that the HR managers focus on upskilling of employees and work persistently on this agenda. The paper is based on a review of secondary sources of data in the form of reports, books and internet sources to conceptually review the technological trends driving the transformations at the organizations today. Survey based reports by Deloitte, McKinsey, PWC etcetera are reviewed to find various technological trends. The paper then raises the concerns of upskilling the employees linking it to the sustainable existence of the organizations. The review also covers the important role of HR in upskilling the employees and various methods to be used to upskill the employees. The paper gives an insight into the best practices being undertaken by such HR leaders of Walmart, JPMorgan Chase, Marriott International, AT&T and Pfizer to upskill their employees and work towards achieving the goal of creating sustainable organizations. The paper concludes that the HR experts need to ensure the implementation of new technologies, upskill the teams, use methods and programs enabling smooth transition and learning of skills amongst the employees. The best practices as adopted by organizations can become a benchmark for others to adopt similar tools and techniques towards upskilling of the workforce.

Keywords: Learning, digital skills, upskilling, sustainability, best practices

## Introduction

The next ten years will see a significant change in how human resources and developing technologies interact, according to recent corporate trends. Use of artificial intelligence, chat bots at workplace, business interactions in meta verse are such few trends leading to a transition of HR's role to a strategic one which is a rapid and profound change. As firms throughout the world undergo a digital transformation, HR's role becomes important in driving that process. How can organizations adapt the HR function to work in a digital manner, leverage digital tools and apps to deliver solutions, and continuously experiment and innovate? Is the current focus of digital HR to bring about organizational sustainability (Deloitte, 2017).

## Technological Trends at the Workplace

According to Deloitte's 2019 Global Human Capital Trends survey, automation, cognitive technologies, and artificial intelligence (AI) are all advancing at an exponential rate (Deloitte, 2019). The organizations are moving towards the following trends as per the report:

- Eighty-six percent of the organizations polled thought that their capacity for learning needed to be reinvented.
- Eighty-four percent of businesses are aware that, in order to increase productivity, they must reevaluate their labor experience, despite the ubiquitous corporate focus on digital transformation.
- And 80 % of respondents think they need to develop leaders differently in the face of new challenges to move more quickly and adapt to a workforce that is much more diverse.

Sustainability of work formulates the biggest concern and towards this end the organizations are working on upskilling their employees and building the competencies required for a seamless transition into the new ways of doing work. The organizations are also revitalizing their talent retention strategies as lucrative reward systems for enhanced motivation amongst the employees. New talent models like contracts, the gig economy, and the Cloud will certainly bring about changes which are unknown to the organizations. Thus, the functions of HR professionals will undergo a drastic change and grow in great dimensions as challenges have increased and responsibilities have changed. The largest challenge for heads of HR is coming from digitalization, said Brian Kropp, Group Vice President, Gartner Research & Advisory.

Twenty thousand years of advancement would occur in 100 years, according to Ray Kurzweil (Chief Engineer, Google), yet in actuality, change has occurred much more quickly than that. Technology is changing society and how people live and work on a daily basis. According to 53 % of respondents, during the next ten years, technology would likely bring about the biggest changes in HR and employment, impacting how we plan, review, communicate, engage, and recruit.HR executives are being encouraged to play a bigger part in pushing the company to "be digital" rather than merely "do digital." As per a survey made by Deloitte in 2017 Deloitte Global Human Capital Trends:

- Fifty six percent of the businesses surveyed reported that they would be rethinking on their HR practices to make better use of digital and mobile solutions.
- Fifty one percent of businesses are currently changing their structures to accommodate digital business models.
- A total of 33 % of the companies surveyed said that the HR teams use artificial intelligence (AI) technology to provide HR solutions, and 41 % are actively developing mobile apps to do so. AI is changing HR practices and its functions, which will be beneficial in lowering human bias, enhancing workforce effectiveness and workplace relationships, facilitating required compliance, increasing metrics adoption, enhancing employee experiences, and fostering workplace learning. Organizations are working to adapt their people and processes to the newest technology after implementing it.

A revolution is about to begin in the field of HR technology. A revolution that will fundamentally alter not only the field of people management but also how businesses and organizations operate as a whole. It has become vitally crucial to comprehend the power of technology and put it to our advantage in today's technologically advanced society. The outdated procedures and slow organizational growth will result from failing to keep up with this technological drift. Social, mobile analytics, and cloud (SMAC) technologies are now used to increase the effectiveness, efficiency, and connectivity of HR. It represents a fundamental change in how human resources operate. Due to the rapid pace of change, HR is faced with a number of difficulties related to the digital disruption. As Julhiet Sterwen noted, some of the difficulties would be (Sterwen, 2019); Making HR procedures more flexible; creating a digital corporate culture; assisting managers with their own digital transformations; reinventing workforce planning; and rethinking the employee experience. According to a Deloitte analysis from 2016, these disruptive developments and technologies include cyber, data, cloud, social, meta and mobile. And a workforce made up of different generations, including millennials or first-generation digital citizens poses a big challenge for HR to manage them in the face of such digital disruptions. These disruptions are also creating new jobs for the workforce. In a report by PWC on Workforce of the Future: The Competing Forces Shaping 2030 it was found that robotics and artificial intelligence might enable the redesign of work, improve productivity and customer experience, and enable a focus on more value-added … And to create these talent streams for future situations predictive analytics and advanced workforce planning can be used. The companies can also work on better employee experiences using this technology (PWC, 2018). McKinsey research from May 2013 listed 12 innovations that will revolutionize business, society, and the economy. These 12 cutting-edge technologies include advanced robotics, autonomous and semi-autonomous vehicles, next-generation genomics, energy storage, 3D printing, advanced materials, advanced gas and oil exploration and recovery, and renewable energy. The report also mentioned several waves to these technological advances. The Internet was the first wave, with improved interfaces being developed and customer support offered via online and mobile devices. The second wave, used AI to improve decision-making, comprehend consumers, concentrate on marketing initiatives, lower expenses, and automate duties. The third wave included the voice, picture, and video recognition. And the fourth wave of autonomous AI is related to automated mobility that includes robotics, self-driving cars, and automated manufacturing processes. These advances thus bring about a greater challenge for HR by means of working consistently on upskilling the employees to maintain viability and sustainability of the organizations.

The objective of this paper is to address these technological challenges and role of HR towards building sustainable organizations by upskilling the employees.

## Organizational Sustainability: Creating Urgent Need to Upskill

Sustainability is the ability to satisfy our needs while not jeopardizing the ability of future generations to satisfy their own needs (World Commission on Environment and Development, 1987).

In a study made by Lange et al. (2012) two empirical papers giving their perspectives on sustainability were studied. And the conclusion drawn out of it was that both sets of research agree that a firm's innovativeness is important for facilitating sustainability (Lange, Busch, & Delgado-Ceballos, 2012). To maintain this innovativeness, focus on continuous upskilling of employees should be rendered here referring to digital upskilling. Sustainable thinking is required for a sustainable future. In a report by the World Economic Forum (Fillmore, 2021) the following observations were made:

- The continuation of the workforce for society is made possible through digital upskilling, which supports sustainability.
- The digital skills gap could widen as a result of the epidemic.
- Upskilling initiatives taken by businesses will pay out financially.

The skill requirement is changing rapidly. In an extensive study made by Asian Development Bank (ADB) and LinkedIn in 2022 to better understand the evolving job market and to emphasize the ascending talents and occupations in the increasingly digital workplaces in Australia, Singapore, the United States, India, Indonesia, Malaysia, and the Philippines it was found that there are three categories of skills required by the workforce as (a) basic digital skills, (b) intermediate digital skills, and (c) advanced disruptive digital skills. The essential knowledge and abilities needed to use enterprise technologies, platforms, and solutions on the job are known as basic digital skills. Those who possess intermediate digital abilities can use hardware and software to create tools, platforms and programs that are simple for anyone with only the most fundamental digital skills to use (Asian Development Bank and LinkedIn, 2022). These can be seen in the Figure 10.1.

How the global economy will adapt to the skills revolution, or the continuing automation of traditional manufacturing and industrial practices, is now one of the major sustainability unanswered questions. Thus, it is clear that providing a sizable number of people with the skills necessary to occupy positions in a new digital economy will soon be essential to the survival and well-being of our society. Because businesses that upskill their personnel digitally are a crucial component of social sustainability, we think it is essential that skill-based initiatives are taken into account in ESG (Environmental, Social and Governance) reporting. Accountability guarantees that the global workforce remains stable rather than on the decline, which is essential for our economy's future (Fillmore, 2021).

According to a PwC report from 2020, businesses with the most sophisticated upskilling initiatives witnessed a threefold increase in innovation. Even more extensive environmental advantages may result from upskilling. Environmental sustainability goals may also be achieved efficiently with an enhanced focus on digital literacy. Human resources specialists have been crucial in helping organizations in re-skilling and upskilling of employees amidst new challenges they face. AIHR (Academy to Innovate HR) has identified some technological challenges which will impact the working patterns of HR as (i) redefining remote and hybrid work strategies, (ii) HR enters the metaverse and (iii) upskilling leaders and managers (Vulpen, 2023).

| Basic | Intermediate | | Advanced |
|---|---|---|---|
| **Basic Digital Literacy and Skills** | **Applied Digital Skills** | **Skills for Computer Software and Hardware** | **Advanced Disruptive Digital Skills** |
| Digital communication and collaboration | Animation | Computer hardware | Aerospace engineering |
| Basic internet skills for commerce and online purchases | Big data use and management | Computer networking | Artificial iIntelligence |
| Mobile banking | Computer graphics | Data storage technologies | Augmented reality and virtual reality |
| Online financial and citizen services | Digital art, music, and videos | Enterprise cloud computing | Autonomous vehicles |
| Social media | Digital data tools | Enterprise digital transformation | Computational thinking |
| Basic computer skills | Digital logistics and warehousing | Game development | Cybersecurity |
| Basic skills to use programs such as Microsoft Office | Digital manufacturing and processing | Information technology administration | Data science |
| | Digital marketing | Mobile application development | Development tools |
| | E-commerce development | Product development | Digital biotechnology |
| | Enterprise software | Scientific computing | Genetic engineering |
| | Financial tech | Signal processing | Human–computer interaction |
| | Graphic design | Software development through programming or coding | Material science |
| | Social media marketing and advertising | Software testing | Nanotechnology |
| | System administration | Network engineering | Platform management |
| | Technical support | Web services and digital solutions | Robotics |

Figure 10.1: Asian Development Bank—LinkedIn digital skills taxonomy

Source: Asian Development Bank and LinkedIn

These challenges have further added to their growing pressure to provide employees with opportunities to advance their skills as our economies become more digital. And this role of HR should be acknowledged and celebrated as a contribution to sustainability when they accept this duty and take steps to upskill their employees (Fillmore, 2021). The consensus among experts is that the workplace is changing quickly. The term "late adopter" in today's environment of rapid change is synonymous with "out of business." Businesses are unable to afford the luxury of waiting and seeing. Talented HR experts will need to stress the importance of and implementation of reforms on all fronts. The effects of upskilling could also be measured by the leaders in four areas as: Employee engagement, Employee retention, Productivity and Profitability, Innovation and Onboarding. Thus, upskilling would just not have an impact on improving the employees edge on their abilities but also work on the various aspects as retaining them, enhancing their productivity and bringing innovations in the organizations.

## Role of HR in Upskilling

In a survey made on five countries(India, Bangladesh, Indonesia, Philippines and USA) by Asian Development bank and LinkedIn it was found that 68 % of the employers are providing digital skills to their current employees. Thirty-eight percent have hired an external expert to provide such trainings and around 51 % of the employees have undergone such trainings (Asian Development Bank and LinkedIn, 2022).

## Upskilling Process to Be Followed by HR (Valamis, 2023)

The HR may accurately create a skill matrix for each employee and ultimately determine the combined abilities and comprehensive knowledge base of the team by employing a methodical methodology.

Investigate the operations and services provided by the companies' primary rivals and determine whether the staff can provide a comparable level of service.

With a better understanding of what the company's personnel are capable of and what is currently lacking, HR can start setting objectives for upcoming upskilling programs.

Upskilling programs can be generic and focus on a certain talent that large segments of the workforce lack, or they can be specialized and made for particular people.

The HR department needs to be proactive and create systems to periodically evaluate team skill sets and identify any gaps that could lower performance for the organization as a whole.

## Methods to Be Followed by HR to Upskill the Employees

Learning and development (L&D) programs, which can deliver training content through many channels, including online or in person, are the

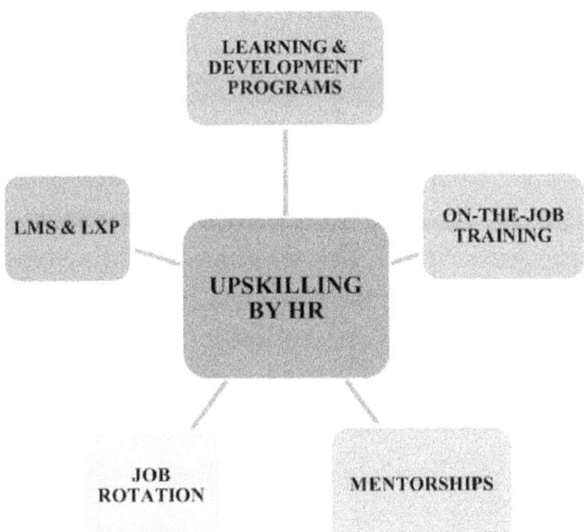


Figure 10.2: Methods of upskilling employees

Source: Self

most obvious upskilling approach. Some of the methods are as shown in the Figure 10.2. L&D programs typically make use of classes, lectures, seminars, and workshops to impart new knowledge to the workforce, making them ideal for large-scale upskilling projects that target numerous personnel.

On-the-job training, a more practical method of imparting new skills, places workers in actual workplaces with qualified specialists. Staff members get the required abilities for later independent applications by performing new activities (under close supervision), as opposed to learning from training materials or presentations. In contrast to more knowledge-based duties, practical or physical work benefits most from on-the-job training.

The best way to pass on knowledge to newly hired employees is through mentorships. The junior team member spends a lot of time learning new skills directly from a senior team member. Employees can swiftly learn new skills, enhancing their capabilities, through direct contact and mentoring.

Employees can take on related positions inside the organization through job rotation or cross-training. It might be difficult for many people to learn through conventional course materials. Through cross-training, students can execute the duties while being closely supervised by knowledgeable employees and expanding their skill set. This assists staff members in learning new skills that will benefit their existing positions or position them for advancement within the organization.

Examples of software to assist organizations in managing their L&D programs include learning management systems (LMS) and learning experience platforms (LXP). LMSs are excellent for managing data regarding a company's training initiatives. This includes providing reports, documentation, and educational materials, as well as enabling HR professionals to monitor the development of various employees. LMSs are also perfect for more general upskilling initiatives aimed at many employees. Contrarily, LXPs strive to offer a customized upskilling experience to each employee in order to get the greatest results. Organizations can create and manage new content based on user choices by using an LXP.LXPs and LMSs complement one another rather than compete with each other. As a result, both enhance the training experience to produce a better informed and more skillful workforce.

## HR Practices towards Upskilling

Some of the best upskilling practices as mentioned in the literature can be adopted by the organizations towards making their services sustainable. Walmart announced in the year 2021 its plan to invest about $1 billion for the following five years to provide its staff with no cost availability of higher education and skill development. It's Live Better U (LBU) initiative entirely covers college tuition and book fees. In contrast to 2018, when it was $1 per day, 160,000 U.S. employees now have free access to it. The company also disclosed four new academic partnerships that will work with the firm's partners in providing courses specifically customized for working professionals (Istenič, Zupan, Česnik, Enija Nosan, & Podpeskar, 2022).

JPMorgan Chase requires new analysts in investment banking and asset management to take Python programming classes because the corporation wants its staff to have a better understanding of the technology so they can collaborate with IT professionals more successfully and provide better solutions. "Coding is not for just tech people, it is for anyone who wants to run a competitive company in the 21st century," the head of JPMorgan Asset Management told the Financial Times (Parkeshi, 2020).

Marriott International has been awarded as the top employer of the Asia Pacific region and offers two staff development programs. The Global Voyage Leadership Development program trains recent university graduates to serve as future leaders within the firm. The 12–18-month program includes hands-on online training in various subjects as accountancy, Culinary arts, Engineering, HR & hotel management. It is nationally recognized. For aspiring managers, the Marriott growth Academy offers a self-paced, blended learning alternative that lets participants take charge of their professional growth. In the Australian market alone, these programs have helped fill more than 55 % of senior positions with internal applicants (GOi, 2022).

Analysis found that just about half of AT&T employees had the STEM abilities that the corporation would require in the future. STEM talents are frequently referred to as the "hard" sciences. The company needed a quick fix to stay competitive in the future and had a few alternatives as hire more employees or retrain and upskill the current workers. With a multi-year, $1 billion plan that includes elements that nearly any company can produce, the company chose to go in the latter direction. The project's main objective is to collaborate with edtech companies to give employees access to online learning opportunities. It has customized learning opportunities on a website which is helpful to the staff to make future career plans and identify the skillset they need to develop (Eightfold.ai, 2021).

At Pfizer, learning is encouraged, not mandated. The Head of Learning and Development at Pfizer stated, We are trying to develop an environment where learning is something that everyone wants to do. With the intention of inspiring and involving employees, the international pharmaceutical firm has developed a particular learning and development program. Learning comes in three flavors: required, desirable, and necessary. Because they operate in a highly regulated and sophisticated sector, Pfizer aims to achieve an equilibrium between required learning (required skills and training) and wished learning. The company pushes its employees to push the limits and learn new skills in order to advance in their careers. Pfizer uses technologies such as learner profiles, competency maps, and peer evaluations to construct their learning journeys (HRForecast, 2022).

There are many more initiatives being practiced across the globe to upskill the employees to meet the digital skill requirement today. The HR should research and practice such interventions in order to maintain the sustainability of their organizations.

## Conclusion

The paper has made a review of the current trends, challenges and the role of HR in upskilling of employees in the vogue of the digital

Figure 10.3: Word cloud of the focused areas in the study

transformation taking place across the globe in the workplaces. The word cloud in Figure 10.3 represents the focus areas in the paper.

The paper suggests that the HR experts need to make sure that implementing new technologies produces fruitful results. To help their organizations deal with issues and recognize the shifting expectations of both businesses and employees. It must be ensured that both the business and employees will gain from the use of technology. The development of AI and automation, along with technological innovation, will have a significant impact on the nature of employment in the upcoming ten years. The HR may anticipate considerable effects on employment and skills as digitization advances. Employees are expected to take on an increasing amount of responsibility for skill development as the workplace becomes more flexible. Acquisition of talent and skills across generations will be crucial. Also, the organizations are currently experiencing the web 3.0, which will completely transform organizations and the future of work. As a result, HR should get ready for future careers, foresee highly in-demand abilities, and keep working towards upskilling and reskilling.

## Editors Note

The exploration of technology's role in upskilling the workforce for creating sustainable organizations, with a focus on the transcending role of HR, holds significant implications for sustainability in the modern workplace and broader society. As technology continues to advance at a rapid pace, organizations are faced with the challenge of ensuring that their workforce possesses the necessary skills and competencies to adapt to evolving job roles and industry demands. By conceptualizing trends in technology-enabled upskilling initiatives and the pivotal role of HR in driving these efforts, this research can offer valuable insights into fostering a culture of continuous learning, innovation, and resilience within organizations. By leveraging technology to facilitate upskilling programs, organizations can empower employees to enhance their capabilities, remain competitive in the job market, and contribute effectively to sustainable business practices. Moreover, by aligning upskilling initiatives with organizational goals and values, HR departments can play a pivotal role in promoting workforce development, talent retention, and overall organizational sustainability. Therefore, this conceptual review has the potential to inform strategic HR practices and organizational policies aimed at building agile, adaptive, and future-ready workforce ecosystems that drive sustainable growth and success in the digital age.

## References

Asian Development Bank and LinkedIn. (2022). Digital jobs and digital skills a shifting landscape in Asia and the Pacific. Retrieved from <https://www.adb.org/publications/digital-jobs-digital-skills-asia-pacific>

Deloitte. (2017). Retrieved from <https://www2.deloitte.com/us/en/insights/focus/human-capital-trends/2017/digital-transformation-inhr.html>

Deloitte. (2019). Retrieved from <https://www2.deloitte.com/content/dam/Deloitte/cz/Documents/human-capital/cz-hc-trends-reinvent-withhuman->

Eightfold.ai. (June 1, 2021). *Successful examples of reskilling and upskilling programs.* Retrieved from <https://eightfold.ai/blog/reskilling-and-upskilling/>

Fillmore, J. (December 15, 2021). Why digital upskilling is a crucial part of sustainability. *World Economic Forum.* Retrieved from <https://www.weforum.org/agenda/2021/12/digital-upskilling-sustainability/>

GOi. (April 21, 2022). *Five companies investing in upskilling the workforce.* Retrieved from <https://goi.mit. edu/2022/04/21/five-companies-investing-in-upskilling-the-workforce/>

HRForecast. (January 5, 2022). *Upskilling program examples.* Retrieved from <https://hrforecast.com/seven upskilling-program-examples/>

Istenič, T., Zupan, N., Česnik, M., Enija Nosan, E. G., & Podpeskar, N. (2022). Reskilling and upskilling in support of company competitiveness. In M. K. Polona Domadenik Muren (Ed.), *METAVERSING the corporate strategy: The opportunities and challenges of digital transformation* (pp. 225–243). Ljubljana: Časnik Finance.

Lange, D. E., Busch, T., & Delgado-Ceballos, J. (2012). Sustaining sustainability in organizations. *J Bus Ethics, 110*, 151–156. <https://doi.org/10.1007/s10 551-012-1425-0>

Parkeshi. (May 3, 2020). *Building technology proficiency in business.* Retrieved from <https://parkershi.com/building-technology-proficiency-in-business/>

PWC. (2018). Retrieved from <https://www.pwc.com/gx/en/services/people-organisation/workforce-of-the-future/workforce-of-the-future-the-compet ing-forces-shaping-2030-pwc.pdf>

SHRM. (n.d.). Retrieved from <https://www.shrmtech.org/overview.html>

Sterwen, J. (2019). Retrieved from <https://www.julhiet-sterwen.com/en/the-hr-challenges-of-digital-transformation/>

Valamis. (March 30, 2023). *Upskilling.* Retrieved from <https://www.valamis.com/hub/upskilling>

Vulpen, E. V. (2023). *11 HR trends for 2023: Seizing the window of opportunity.* Retrieved from <https://www.aihr.com/blog/hr-trends/>

World Commission on environment and Development. (1987). *Our common future.* Oxford: Oxford University Press.

# List of Contributors

## Part I: ICT and Business Problems

CHAPTER 1 Examining the Mediating Role of Patients' Satisfaction and Hospital Reputation on the Influence of Corporate Social Responsibility on Patients' Loyalty in Ghanaian Public Hospitals

Tarik ATAN, Japheth Nuhu Ahmed and Dominic J. Kamara: Department of Business Administration, Cyprus International University, Nicosia, Cyprus

Dennis Affum Osei: Department of Marketing Management and Business Strategy, University of Hull, Hull, UK

CHAPTER 2 Impact of Total Quality Management on Revisit Intention through the Mediating Role of Patient Satisfaction on Inpatients in Public Hospitals in Addis Ababa, Ethiopia

Dagnu Haile Tesssema, Dr. Figen Yeşilada, Hayford Asare Obeng, Japheth Nuhu Ahmed and Dominic J. Kamara: Department of Business Administration, Cyprus International University, Nicosia, Cyprus

CHAPTER 3 A Scientometric Analysis of Swarm-based Metaheuristics Approach

Samuel-Soma M. Ajibade: Department of Computer Engineering, Istanbul Ticaret Universitesi, Istanbul, Turkey

Oluwadare Joshua Oyebode: Department of Civil and Environmental Engineering, Afe Babalola University, Ado Ekiti State, Nigeria

Sushovan Chaudhury: Department of Computer Science and Engineering, University of Engineering and Management, Kolkata, India

Anthonia Oluwatosin Adediran: Department of Estate Management, The Federal Polytechnic, Ado Ekiti, Nigeria

CHAPTER 4 The Mediating Effect of Patient Trust on the Link between Relationship Marketing and Hospital Brand Image in Public Hospitals-Nigeria

Japheth Nuhu Ahmed, (Dr.) Figen YEŞİLADA, Hayford Asare Obeng, Dagnu Haile Tessema, Dominic J. Kamara, Dennis Affum Osei and Peter C. Mornyey: Department of Business Administration, Cyprus International University, Nicosia, Cyprus

CHAPTER 5 The Effect of Organizational Culture on Individual Creativity: An Empirical Investigation of Packaging Firms in Büsan Industrial Site of Central Anatolia Province, Konya-Türkiye

Aleem ABDUL-KAREEM: Faculty of Economics and Administrative Sciences, Department of Business Administration, Selçuk University, Selçuklu-Konya, Türkiye

Stephen Taiwo ONIFADE: Faculty of Economics and Administrative Sciences, Department of International Trade and Logistics KTO Karatay University, Konya, Türkiye

Part II: ICT, Energy and Financial Sustainability

CHAPTER 6 The Impact of Credit Risk on Bank's Performance after the World Financial Crises: Evidence from 5 Commercial Bank in Sierra Leone

Jeneba Joy Tucker: Department of Business Administration, Cyprus International University, Nicosia, Cyprus

Dr. Murad Abdurahman Bein: Faculty of Economics and Administrative Science Cyprus International University, Nicosia, Cyprus

CHAPTER 7 Analyzes of Common Behavioral Bias among Indo-Arabian Investors and How to Better Manage Them

Richard Sagor Mitra and Dr. Hemant Bhanawat: Amity University, Jaipur, India

CHAPTER 8 Determinants of Pensions: Empirical Evidence from OECD Countries

Akash Kumar Biswal and Biswajit Patra: Department of Economic Sciences, Indian Institute of Science Education and Research (IISER), Bhopal, India

CHAPTER 9 Associability between Overconfidence Bias and Financial Risk Tolerance with Reference to North India

Dr. Sushma: IMS, Lucknow University, Lucknow, India

Dr. Anurag Shukla: School of Management, Sir Padampat Singhania University, Udaipur, India

CHAPTER 10 Technology Enforcing Upskilling of Workforce for Creating Sustainable Organizations: A Conceptual Review of Trends and Transcending Role of HR

Dr. Shweta Lalwani: School of Management, Sir Padampat Singhania University, Udaipur, India

# Index

behavioral bias 155–157, 159, 161, 168, 202, 203, 214
behavioral finance 155–158, 164, 203
best practices 197, 217
bootstrapped panel quintile regression 173, 176, 177
brand image 69–76, 78, 80–87

corporate social responsibility 3, 4, 6, 8, 9, 12, 13, 15, 20, 22–29, 47, 90, 152
creativity 93–95, 98–103, 105–111
culture 47, 93–98, 100–103, 105–111, 160, 168, 169, 214, 220, 230

digital skills 217, 221–224

GDP per capita 173, 180–183, 186–188, 192, 193, 196

healthcare institutions 3, 5, 9, 10, 20, 21, 22, 33, 47, 69, 71
hospital reputation 3, 6, 9, 10, 11, 12, 15, 16, 18, 21, 22
hospitals 3, 5, 9, 10, 11, 15, 16, 18–22, 31, 37, 45–47, 69–71, 76, 83–87

Indian Stock Market 155, 159, 205
individual creativity 93, 94, 98–103, 105–111
investment decision 155, 156, 158, 162, 166, 168, 201, 202, 208

learning 59, 214, 217–219, 225–228, 230

metaheuristics 64, 65

non-performing loans 119, 120, 122, 126, 130, 131, 134, 139, 141, 143–146

OECD 173–178, 180, 182–187, 191, 192, 194–197
optimization 51–53, 55, 59, 62, 64, 65, 111
organizational culture 93–103, 105–111
overconfidence bias 164, 165, 201, 205–208, 210, 211, 212–214

particle swarm 51–54, 58, 59, 61, 64
patient loyalty 3, 11, 16, 21, 32, 35, 49, 50, 90
patient satisfaction 6, 11, 15, 20, 21, 23, 29, 31, 32, 33, 34, 35, 36, 37, 38, 40, 43, 45–50, 87
pension system 173–181, 183, 184, 189, 194, 195

relationship marketing 70–76, 78, 80–87
return on assets 119, 125, 129, 139, 144
revisit intention 31–33, 35–40, 43, 46, 47
risk-tolerance scale 201

Saudi Stock Market 155, 162
scientometric analysis 51, 64
swarm-based 51, 64, 65

total deposit 119, 131, 134

upskilling 217, 218, 221, 222–230

word-of-mouth 35, 38, 40, 42, 50, 71, 84

www.ingramcontent.com/pod-product-compliance
Ingram Content Group UK Ltd.
Pitfield, Milton Keynes, MK11 3LW, UK
UKHW021844100926
8556IPUK00019B/1048

9 781803 745763